# The Best of M.J.C.
# The Weirdness Continues!

by Martin J. Clemens

1st (Print) Edition

Copyright 2015 Martin J. Clemens
ISBN-13: 978-1511745895
ISBN-10: 1511745894

Contents

# Contents

# Prologue

The following essays originally appeared in their entirety on martinjclemens.com. They have been collected here to provide a Best-Of compendium of the written works of Martin J. Clemens. If you enjoy this collection, please visit the website for more from Martin and from other guest contributors.

In addition, further writings by Martin J. Clemens can be found at mysteriousuniverse.org, dailygrail.com, and ancient-origins.com, as well as in the Darklore anthology - volume 8, from Daily Grail Publishing.

If you would like to connect with Martin on social media, he can be found at the following:

http://twitter.com/ForteanWriter

https://www.facebook.com/martinjclemens

Martin J. Clemens is an author and blogger in the realms of science, history, Forteana, and mythology. His works cover topics ranging from Artificial Intelligence and genetics, to archaeology and anthropology, as well as topics than can only be described as...weird. A background in law enforcement, investigation, and research affords Martin an array of experience and skills that provide a unique view of most topics, and the ability to distil complex subjects into easy to understand commentary with avenues for further learning.

# Science Stuff

## Epigenetics and Cellular Memory: Where Are We Headed?

Do you remember your high school biology classes?  Yes, the one where you dissected the frog.  Do you remember anything besides the frog?

If you do, you should have some basic understanding of what gestation is: which is the cellular development process that happens immediately after an egg is fertilized by the winning sperm cell, or the joining of both male and female gametes.  Once fertilized, the egg or ovum quickly (in relative terms) becomes a zygote.  The zygote is, if you'll recall, a single celled version of the parent species, but only for a brief moment.  Very soon after the formation of the zygote a process called cleavage happens and thus begins the everlasting process of cell division or mitosis.

This process never stops, either.  It's happening inside you right now, and will till the day you – and they – shed this mortal coil.

Having said all that, are you aware that there are over 200 different types of cells in your body?  Skin cells, liver cells, and brain cells for example.  How do you think each cell figures out what type of cell it's supposed to become when it splits?  You could easily say that a lung tissue cell will produce two more lung tissue cells, as they simply copy the structure and genome expression of the parent, but there's actually a little more to it than that.

There's a lot of information contained in DNA (somewhere in the neighbourhood of 2GB), and there's a lot of

DNA in every cell, but as certain preeminent biologists have said, DNA is only the recipe, we still need to understand who or what the cook might be. We're pretty damn close to an answer, too. Which in and of itself needs explaining.

While the mechanics of mitosis and the potency of various cell types (possessing the ability to transform into many different kinds of cell) are quite well known, only recently have scientists been able to study the ways in which cells pass on identity information from generation to generation. This is known as cellular memory, or epigenetics, and as a field of study it's advancing by leaps and bounds at the moment.

It's sort of a complicated thing to explain. It involves transcription factors, or basically, an appendix to the book that is your DNA, that are arranged into a cellular transcription profile. They are, essentially, a specialised group of proteins that act as instructors to all the other necessary proteins in your cells. It is transcription profiles that are responsible for informing cells (and the proteins that make up cells) as to what they're supposed to be, in comparison to the cells around them, and instructing them on how to go about fulfilling that role. Sort of a DNA user manual.

As I recently wrote, the study of epigenetics has proven that these transcription factors function as an actual cellular memory. The information contained in these proteins holds not only instructions for the attributes of other cell types, but also an historical record of the identity of the line of cells that came before it; a cellular family tree, so to speak.

These things are now fact – though study continues – but on an even more mysterious level, these transcription factors also seem to pass memories (or some form of memories) from the host organism's experience between individuals through birth and even organ transplantation.

This is all well and good, debatable as it is, but this is a completely natural process. It happens in every living thing, constantly, and we evolved to be this way. However, researchers are now able to artificially induce epigenetic expression of individual transcription factors, and this is an exciting development.

This is something scientists have been doing with stem cells for quite some time, but there are differences in technique. Stem cells have a high potency, which means that they are basically blank slates waiting to be told what to become. They are capable of becoming nearly any other type of cell and need a relatively simple chemical signal to send them on their way. But there are many different kinds of cells that can do this.

There are two main methods for what's now called synthetic epigenetics: somatic cell nuclear transfer, which is a genome wide change in the epigenetic state of the cell; and induced pluripotency, which is the selective expression of specific groups of transcription factors. Both of those approaches are termed top-down methods, in that they are only indirectly affecting the transcription profiles of cells, allowing researchers to study the effects, rather than the mechanics.

Researchers from the University of Stuttgart in Germany have published a detailed review of what they call a bottom-up approach to synthetic epigenetic manipulation, citing the current methods, procedures, and technologies used in this endeavour.[1] A bottom-up approach is different from the traditional methods because it allows a much more targeted and direct interaction with the mechanisms of epigenetic transfer, and the ability to observe and interact with specific epigenetic markers like never before.

The authors of the paper suggest that this more direct approach is poised to provide answers about genetic diseases, such as cancer; specifically what epigenetic markers are responsible for passing the inherited predisposition for such disease.  This would be an incredible break though, but the alarmist in me is concerned about where this could lead.

If, or perhaps the better word is when, researchers finally gain enough insight and technical skill to eliminate a disease such as cancer from the human genome through the manipulation of cellular transcription profiles, they will also have gained the ability to change other aspects of our genome in a much more controlled manner than is currently envisioned. They would also be able, in theory, to add transcription markers to our DNA, thereby creating new genetic traits at will, or, in the case of cellular memory, perhaps implant knowledge, experience, and memories into an infant during gestation, resulting in babies being born with upgrades.  The sky seems to be the limit, as it were.

That would be a long way off, if it is indeed possible, though it does seem to at least be plausible based on current research.  The only thing standing in their way is the technology, which – as the above mentioned researchers noted in their review – is steadily outpacing research right now.

As we plough ahead, lapping up the knowledge held in our very essence, are we learning how to build a better us?  Are we on the doorstep to a new world of genetically perfect humans, or is this just yet another road to eugenics, discrimination, and the polarisation of our society?

[1] Tomasz P Jurkowski, Mirunalini Ravichandran, and Peter Stepper. Synthetic epigenetics—towards intelligent control of epigenetic states and cell identity. Clinical Epigenetics March 4, 2015, 7:18  doi:10.1186/s13148-015-0044-x

# Spontaneous Nuclear Fission: How Mother Nature Made Bombs

So, you know what a nuclear bomb is, right? Of course you do. Everyone does. The very idea is ubiquitous throughout human culture. Do you know how a nuclear bomb works? That's another matter entirely, isn't it?

It's actually not that complicated. Well, the basic concept isn't. Getting one to explode is another story. You've heard the term "splitting the atom" right? That was the big breakthrough proposed theoretically in 1938 by several scientists, then experimentally confirmed in 1939 by Otto Frisch, and then eventually realised as a weapon by The Manhattan Project back in 1945, forever changing the world and its prospects for survival.

Essentially, what happens when you split an atom – meaning the breaking open of the nucleus of an atom (hence the term nuclear) – is that an enormous amount of energy is released. Which, obviously, is quite handy when your goal is death and destruction.

When we refer to "splitting the atom", we're not talking about taking a really tiny knife and chopping an atom in two. The term refers to an induced fission reaction. Fission is the name for the entire process, and in this case it begins with neutron bombardment. What happens is, an atom, uranium-235 for instance, is struck with a tiny particle called a neutron, like they do in particle accelerators such as the Large Hadron Collider at CERN. That collision causes the uranium-235 to briefly change into uranium-236 (a slightly heavier element). The excitation of the particles caused by the collision in turn causes the atom to split apart, releasing the faster-moving, lighter elements within (fission products), as well as the free neutrons it contains. Those particles, once separated, spread out from the nucleus at extremely high-energy, which is what provides the power...or destruction.

So, as mentioned, the process is relatively simple, but building and detonating a nuclear bomb is not something one can do accidentally. It takes a great deal of technical knowledge, equipment, resources, and money.

It's a slightly different story for a fission reactor, though. The kinetic energy of the fission products in a bomb are allowed to continue expanding, releasing that energy directly into the environment (and destroying anything in the vicinity), but in a reactor that energy, along with the electromagnetic energy from the gamma ray bursts is converted into heat energy when the particles collide with the atoms of the specially constructed reactor material and its active fluid (usually water). That heat energy is then used to power steam turbines to produce electricity.

You'll notice the distinct lack of explosions related to fission reactors, with some pointed exceptions. But as with bombs, the process requires an initial input of energy to get the reaction started. You might wonder though, as I did, why this doesn't happen naturally, and the short answer is...it does.

The Sun, of course, is actually a giant nuclear fusion reactor. It does the same thing any nuclear power reactor on Earth does, except it doesn't split the atoms, it fuses two or more smaller atoms together by way of gravitational pressure (which is the difference between fission and fusion). But the Sun, and its many billions and billions of star cousins, aren't the only places these reactions happen spontaneously.

They've happened here on Earth too, a very, very long time ago. The continent of Africa still holds some secrets, and one of them resides at Oklo in the Central African state of Gabon. Oklo was at one time Europe's richest source of energy grade uranium. France operated several mines in the region for decades, until eventually the supply was exhausted. The uranium mined at Oklo has been used in nearly every nuclear reactor in Europe, but in 1972 it was discovered that there was something odd about some of the samples.

French physicist Francis Perrin found that comparisons of uranium-235 from several sites brought up clear discrepancies. Uranium ore, like any ore, is found in different places all over the world, and the ore from each location has subtle differences that can be used to trace the origin of the enriched ore later in production. Normally, Oklo's ore had a uranium-235 content of 0.7202%, but some samples were found to contain a concentration of only 0.600%, which is a considerable difference. It was almost as though the uranium had already been inside a reactor...it was depleted as though it was already used.

Upon further study it was found, quite remarkably, that the uranium deposits at Oklo, in some sixteen different locations, had undergone spontaneous fission inside a naturally occurring reactor approximately 1.7 billion years ago. Special conditions were indeed needed for this to occur. It turns out that the ratio of 235U to 238U just happened to be nearly the same as is used in today's nuclear power reactors, and all that was needed was for the introduction of ground water to act as a neutron moderator and voila! You've got yourself a spontaneous fission reactor.

Despite the fact that the Oklo reactors are the only ones that have ever existed on Earth, we actually know a great deal about when they occurred and for how long. It's been determined, by studying the decay products of the reaction – such as the amount of xenon in the surrounding material – that each reactor was active on a continuous cycle of thirty minutes of criticality followed by two hours and thirty minutes of cool down.  And this went on for several hundred thousand years.

Interestingly, some believe that a spontaneous fission reaction such as those at Oklo, which may have taken the form of a detonation rather than a reactor, may have been the explosion event that created Earth's moon.  While not a widely accepted hypothesis, the idea does have some evidentiary support, though so does its competition, the giant impact hypothesis.

It's also thought that Mars, which is thought to have an abundance of radioactive uranium, thorium, and potassium, might have experienced these types of localised spontaneous fission reactors as well, in the very distant past.  Which may offer future Mars explorers a source of energy on the red planet, if we ever do get there.

So now you know how a nuclear bomb works, and how nuclear physics, with is complicated math and confusing terminology is actually fairly simple, when you get right down to it.  It's so simple, in fact, that Mother Nature can do it all on her own.

# The Smart People Are Worried About Artificial Intelligence, Are You?

2014 was a big year in science. Well, it was a big year in other respects too, but several scientific fields reported some pretty impressive achievements over the past 12 months. The next 12 look to be quite impressive as well.

One of those achievements was the long pursued realisation of Isaac Asimov's dream: Artificial Intelligence.

That's a loaded sentence, of course. There are a few caveats that go with it. As you'll read in that previous post, the researchers from the OpenWorm Project mapped the brain – that is the neural connections of the brain – of a nematode. They then recreated those connections in a virtual framework and connected the result to a Lego-robot. And lo and behold, it worked! The details are fascinating, as you well imagine, but the point is that we achieved AI...for better or worse.

True to form as human beings though, we've leaped over the gap before we looked at what was on the other side. Though not before we were warned by pre-eminent scientists and Hollywood producers that such a leap could be fatal.

Back in May of 2014 renowned theoretical physicist (and cyborg) Steven Hawking got together with some pals (Stuart Russell, Max Tegmark, and Frank Wilczek) and the four of them sent us all a sobering missive in the form of an OpEd piece in The Independent. OK, maybe it was more of a request for dialogue, but either way, they broached the subject of how dangerous Artificial Intelligence could actually be in the real world.

We heard their message loud and clear. Or well...actually we completely misconstrued their message (and freaked out a little) over what was supposed to be a call for discussion. Then it all died down and went away. Until Hawking sent us all another memo in December, but this time he was, shall we say, less charitable with the subject. His comments in this instance were in response to questions about his new Intel-developed communications system which allows him to speak in-spite of being stricken with ALS. That system being a simplified form of AI.

He was quoted as saying that "Artificial Intelligence could spell the end of the human race", which is decidedly less ambiguous than his previous comments. Of course, even that was misconstrued. Some outlets claimed he was denouncing the development of AI entirely, or that he's stocking supplies for the robo-pocalypse. But through it all, his overarching point has been that we need, collectively, to talk about where we're going with all of this.

To further that point, Hawking has joined a group of what some might call the last best hope for mankind. The Future of Life Institute (FLI), founded by an all-star group of scientists and tech experts, including Jaan Tallinn (the cofounder of Skype) and Max Tegmark (world renowned cosmologist) has issued an open-letter to the AI technology development world and the greater community of policy makers, calling for deliberate and comprehensive research and discussion regarding the safety of AI.

The FLI's scientific advisory board looks like a who's-who of smart people, including Steven Hawking, Elon Musk, George Church, Alan Guth, Cristof Kock, Saul Perlmutter, Martin Reese, and even Morgan Freeman and Allan Alda, among others. The brain power behind this initiative is almost as scary as what they think of AI.

Let's be clear here; what they're calling for is both reasonable and prudent. Such deliberate consideration should must be undertaken prior to any scientific or technological advancement. The pros and cons should be weighed carefully and by the best minds we can muster. But (you knew there would be a but) fear-mongering here is no different than fear-mongering on cable news.

We each have our roles to play; the scientific elite have ethical problems to solve, as well as technical ones. Our philosophic elite have moral questions to ask (and try to answer).  And we, that is you and me, have an obligation to understand what is being said in those two forums.  It's not enough to read a watered-down news article on the Huffington Post, telling us that the smart people are afraid of AI and we should be too, we have to go to the source to find out what those smart people are actually saying.

In this case, those watered-down articles seem to want to demonise the coming robot overlords as a collective entity bent on our destruction simply because we exist.  And while that might make for a good movie, it's hardly realistic.  But who could blame you for thinking when Steven Hawking says "artificial Intelligence could spell the end of humanity" that he means Skynet and its Terminators are on their way.  He doesn't, mind you.

There's a concept in AI culture that many people would benefit from knowing well.  It's called Grey Goo.  This unfortunately named idea says, quite simply, that self-replicating robots (whether self-aware or not) could decimate the face of the Earth in a matter of days.  That may sound to you like more fear-mongering, but it's a very real threat.  How's that, you ask?

The concept of Grey Goo is also known as ecophagy (eating the environment), and with that you can see what the danger is.  It's not an anthropomorphic warning that robots will declare us the enemy and fight passionately for our destruction. It says that the robots will do what they were meant to do, but they'll do it too well.  They consume resources to replicate, and as they develop, the speed at which they do it increases exponentially.  The point is, they don't stop...ever.  And we become the material that they use to replicate.

There's no malice, there's no war, there's no declaration of technological superiority...they just do what they do.

Hawking, et al, are suggesting quite the same thing, except that Grey Goo is meant to describe nanotechnology.  The fear is that, once higher order artificial intelligence is achieved and unleashed, its self-directed advancement would essentially render humanity (or even biological life) obsolete. And what do you do with obsolete technology like old cell phones?

The popular discussions about this, specifically on Hawking's comments, revolve around those more anthropomorphised fears. They are the result of us projecting a human psyche onto something that is most definitely not human. And that isn't what FLI and its proponents are warning, however that doesn't mean we're completely wrong to discuss it.

As you'll note from the post mentioned above (about robot worms) the success of that project was due, in part, to how simple the nematode brain is, but higher order intelligence is wholly different. Where those researchers were able to transfer basic motor control and stimulus response to the robot, if later researchers can do the same and more with more complex brains, we don't know what can be transferred. Instincts? Memories? Personality? Human nature?

Will it be similar to booting up a computer with the most basic operating system installed? Wherein that intelligence will just do what it does? Or will it be like switching on the internet? Complete with bad attitudes, conflicting opinions, and a general hatred for mankind?

You can see, or you should be able to see, that the warnings we've gotten from FLI and Hawking are not about the latter. But even so, they are just as (potentially) dire.

Here's the rub though...We're well on the way to realising the version of AI that's represented by the Grey Goo idea, and when we consider the pitfalls on that road, we have to strenuously resist the urge to project our own psychology onto that technology. There may come a day though, when we can answer some of those questions about transferring humanity into a machine, and we must then consider how the robots might feel about us.

But let's deal with one at a time.

# Extended Musings on Determinism and the Origin of Consciousness

I'm reading a book. It's a book written by a friend. I call him a friend even though I've not met him, nor, likely, will I ever meet him. That isn't why I'm reading it though. I'm reading it because it covers a topic that is dear to me, a topic that has vexed me for many, many years: Death.

No, I'm not a brooding and morbid existentialist. Actually, now that I think about it, yes I am, but that's neither here nor there. It isn't the spectre of death that interests me. It isn't some dark obsession with evil or gore or fear that has me vexed. It is, as I think my previous writings have demonstrated, the mind-body question that has my attention. It is the divide between the deterministic world view and the spiritualist world view. It's an intellectual pursuit driven by a single question: what's next?

I mentioned the book I'm reading, which is Greg Taylor's *Stop Worrying! There Probably is an Afterlife* (Daily Grail Publishing 2013), because even though I really just started it, my friend has gotten me to thinking – not a terribly difficult thing to achieve, I'll admit. The second chapter is dedicated to one particular aspect of the mind-body question, or more specifically the question of "what's next?" It deals with the idea of death-bed visions.

If you are at all familiar with my writings on this subject, you'll probably already know my take on these matters, but I will say that Greg's thoughts on the subject have affected my view. It should be known that I intend to provide an in-depth review the book upon finishing it, what follows, however, is not a review, but rather is my take on one aspect of its contents.

Death-bed visions are, much like near-death experiences, almost archetypal to the process of dying. At first glance they seem like an esoteric experience based in delusion, or hallucination, but upon further reading, one inevitably finds that not only are death-bed visions far more common than most anyone realises, but they are surprisingly similar among accounts. There are literally thousands of documented death-bed vision experiences, and, even by the most conservative estimates, millions of undocumented experiences. Depending on who you ask, these experiences seem nearly ubiquitous.

I will be dedicating an entire post (perhaps two) to the topic of death-bed visions, as it holds much potential to understand what happens to us when/as we die, for now though, I'd like to focus on an issue that, according to some, predicates all other issues on the topic of the mind-body question: Determinism.

Don't run away just yet! I know, determinism is a dirty word. I know that the philosophy of life is one of the topics our betters have warned us not to discuss in polite company, amongst other subjects like politics and religion, but come on...take a risk with me.

Determinism says that acts of the will, occurrences in nature, or social or psychological phenomena are causally determined by preceding events or natural laws. Which means, in simpler terms, that there is a physical cause or origin, rooted in nature, for all aspects of reality.

Greg Taylor, a wiser man among many wise men, has asserted in his book that a deterministic world view is flawed. He says that it's inadequate, and that experiences like death-bed visions are evidence that determinism is an infantile pet concept of aggressive atheists and modern philosophers of mind.

It's true, many so-called angry atheists subscribe to the deterministic view of biology. P.Z. Meyers is quoted as saying: "If many object to the idea that human identity emerges gradually during development, they're most definitely going to find the idea of soullessness and mind as a by-product of nervous activity horrifying. This will be our coming challenge: to accommodate a view of ourselves and our place in the universe that isn't encumbered by falsehoods and trivialising myths."

Richard Dawkins has made similar statements, as have several other popular atheists in the limelight. Meyer's words embody the argument, but don't really counter it. He seems to dismiss, or conveniently omit one very important aspect, one that features prominently in Greg Taylor's argument, and that is that consciousness appears to be pointedly non-deterministic.

This position is valid. It's demonstrated in the phenomena associated with the afterlife, though in the interest of intellectual honesty, that phenomena isn't readily accepted by those who don't subscribe to any kind of religious tradition. It is largely unquantifiable, which is precisely why it sits in opposition to determinism. That doesn't mean it should be dismissed, it means only that it should be recognised as less valuable to the discussion than some would prefer. It doesn't end there though.

A non-deterministic view of consciousness is also supported through quantum physics. Some might suggest that this is no more measurable than afterlife phenomena, and to those unfamiliar with the complex ideas and theories held in quantum mechanics, that might be true, but concepts like entanglement, uncertainty, and even zero point fields are quite real and measurable.

But we needn't go that far in our pursuit of ways to quantify consciousness.

I recently wrote about some (relatively) new and exciting brain scanning techniques emerging from the field of biomedicine in neuroscience. In an unrelated post, I discussed the possibility that geomagnetic energy might be affecting our brains and causing hallucinations of a sort; hallucinations that are in turn perceived as paranormal experiences. As a result of that post, and the commentary that followed, I came to realise that the evidence for determinism is now stronger than ever, and the reason is much simpler than quantum physics.

It's true that there really is no disconnection between people. In fact, there is no real disconnection between anything that exists. All matter – and energy – are made of up fundamental particles. This is true for what we would call solid sovereign objects; they exist in relative separateness from all other objects, but this is an illusion.  It's easy to visualise the connectedness of objects that exist inside a medium like water, wherein the idea that the water acts to connect objects is somewhat intuitive.  The idea, of course, is that the molecules, the atoms that make up the water, are in constant contact with the molecules and atoms of the objects in question, thereby forming an unbroken – and some might say unbreakable – chain between object, medium and object.

The same is true for any object that exists in any medium, such as the atmosphere of the earth for instance. What we call air is no different than water, in that it is made up of fundamental particles; molecules and atoms, and as a result, those things that exist within it are fundamentally connected to any and all other objects that exist within the same medium.

But are we talking about anything more than just atomic proximity?  The determinist would probably argue no, we are not; while the dualist might argue the opposite.  Myself, I side with the dualist; standing next to a light post, no more makes me a light post than it makes the light post me. Proximity alone is not enough confer connectedness in the sense offered by dualists.

Interaction is the key factor in demonstrating the connectedness of two (or more) objects or entities.  This is embodied in my geomagnetically induced hallucinations post. As I mentioned in that post, the new brain scanning technology called transcranial electrical stimulation, or TES, and its variants, uses electrical impulses to affect specific areas of the brain as may be targeted for medical or research purposes.  Those effects are varied and of themselves are quite interesting, but what's important here, is the fact that an outside device can affect the cognitive function of a human brain.  It does this by projecting electromagnetic energy into the brain in finely tuned ways and frequencies.  In fact, as this technology develops, in conjunction with other technologies and techniques, researchers are able to more finely target specific brain regions, allowing them to pin point and map what regions of the brain are responsible for what cognitive or behaviour processes, on an almost one-to-one scale.

It may not need to be said at this point, but the very fact that this can be done, stands as the strongest evidence yet that cognition – that is thoughts, feelings, motives, etc. – are the product of nothing more than the biomechanical nature of the brain. It says that consciousness is not only housed directly in the brain, but also originates there.

Obviously, this idea is diametrically opposed to the dualism of mind philosophy, and some of its proponents would argue, as an analogy, that my television depends on both its internal components and the signal provided by my cable company in order to work properly. Disrupting either will cause problems.

I agree, that seems reasonable, but our brains are not TV's.

The false analogy between televisions and brains offers an inaccurate picture of the issue. Firstly, a TV has an obvious connection to the signal provider – a cable, or in some cases a wireless transceiver – but our brains do not. Mammalian brains, or even lower order brains for that matter, have no apparatus for sending or receiving signals of any kind. In the case of a television, if you physically disconnect the source of the signal, the set will not work (properly). You can even go so far as to enclose the TV in a Faraday cage, and the effect will be that no signal of any kind can reach its transceiver, thus it will not work.

This isn't so in the case of a human brain. A person can be isolated in every way we can imagine, and yet their brain continues to function; we see no interruption in consciousness whatsoever. If you were to place a human inside a Faraday cage – which will block any electromagnetic signal or field, depending on the size of the mesh – there is no effect on consciousness.

You might argue that there are fundamental energies that we don't yet understand, so it's possible that such energies require receiving/transmitting apparatus that we might not recognise. It's also possible that whatever signal or energy might be responsible for consciousness isn't electromagnetic in nature, thus wouldn't be affected by shielding.  However, as I'm sure you're aware, there are only four fundamental forces (read energies) in the universe: electromagnetism, gravity, and the strong and weak nuclear forces.  So there is little else to work with, even when you factor in the mysterious zero point energy, the existence of which is as yet still only an untested hypothesis.

I can already hear the argumentative machinations of the dualists winding up.  "But there could be an energy that we just haven't discovered yet!"  Sure, there could, but we've seen zero evidence of such an energy in all other endeavours.  Nature isn't that wasteful.  The four fundamental forces are found everywhere.  They have an effect on everything...all four of them.  How could it be that there is another energy that only affects consciousness and which doesn't interact with the other four in any way?

There are other counterarguments though, such as the universality of near-death experiences and the related imagery, not to mention death-bed visions as mentioned above.  Though, the outside influence of a controlling or supplying energy isn't required to explain those phenomena.  Even if you dismiss the delusion/hallucination explanations typically offered by those pesky determinists, there is still room for explanations for those experiences that doesn't invoke mysterious new energies, magic, or omnipotent deities.

For the record, I'm still on the fence, which is a position I wish more people would take up.  This issue, among all others, is the one issue that does not warrant certainty from either side.

# Was Oliver the Humanzee a Missing Link in our Evolution?

There's a lot of confusion these days regarding the concept of evolution. In general terms it's a fairly simple process, but the details can be a little harder to grasp. There are elements of the concept that are almost infamous for their wide misrepresentation by certain segments of society. One of those is the idea of a transitional species.

That terms refers to individual species that represent the evolutionary path from an older one to a newer one. The confusion about this idea is embodied by the infamous missing link argument. That argument originates with ancient Greece and the great chain of life concept, which extended into modern Neoplatonism. Later adopted by the deist school of thought regarding the origins of life, it details a hierarchical structure to all life as decreed by God. The progression is given as starting with God and proceeding down through angels, demons (fallen/renegade angels), stars, moon, kings, princes, nobles, men, wild animals, domesticated animals, trees, other plants, precious stones, precious metals, and other minerals. The missing link is therefore any species we would expect to see bridging any gap between any one of those categories, but which is not represented in the record.

This mindset, though pagan in origin, has dominated religious thought for centuries, and with the monumental breakthrough that was Charles Darwin's On the Origin of Species, the primacy of the idea began to break up. The missing link idea is now one of the most oft-leveled arguments against the validity of the Theory of Evolution.

Popular culture has exploited this confusion for its own purposes many times. Virtually every monstrosity that has graced the silver screen has been an example of the missing link in one form or another (at least those monsters that originated on Earth). King Kong, the Planet of the Apes, even Paramount

Pictures' Monsters vs. Aliens with their namesake character, the Missing Link, have cashed in on the idea for much gain. And who could leave Bigfoot or Sasquatch out of this discussion? The popularity of the missing link idea seems to have accelerated a general misunderstanding of the process of evolution and there have been a few real life examples of creatures that were thought to be missing links in our evolutionary past.

Archaeopteryx, Australopithecus afarensis, and the tiktaalik are famous examples of purported missing links in the fossil record. In each case it's been argued that the fossil showed transitional characteristics between a parent species and a child species, though each has been debunked, so to speak.

A lesser known and far more recent example of a purported transitional species, personified in a living creature, was Oliver the Humanzee. As his name might suggest, Oliver was a chimpanzee. He was found and taken from the Democratic Republic of the Congo in 1960 by primate trainers Frank and Janet Berger. The Berger's became interested in Oliver, then approximately two years old, because his behaviour and appearance suggested to them that he was a different species of primate than his peers, one much closer to humans. Oliver's face was flatter than other chimpanzees and has human-looking characteristics, and he almost exclusively walked upright. It was said that he preferred the company of humans over other primates and mimicked human behaviour whenever possible.

Oliver lived with the Berger's for the next 16 years, when in 1977 he was handed over to a small theme park in Buena Park, California. It seems that Oliver had taken a romantic interest in Mrs. Berger and had become a danger to her after several attempts to mate with her. The next 21 years of his life had Oliver being showcased as an oddity of the natural world and used as a test subject for scientific and cosmetic research, until he was retired, so to speak, with the primate sanctuary Primarily Primates in Texas in 1998. At that point he was largely blind and terribly arthritic, but conservationists and rehabilitators provided him with a comfortable existence until his death in 2012.

As mentioned, Oliver was, for a period of more than 20 years, considered to be a living example of a transitional species or missing link between apes and humans. Many people speculated that he was the first in a line of primates to evolve beyond being a mere animal and could have been the progenitor of an entirely new species of proto-human.

There are many problems with this idea though, not the least of which is the fact that all species are transitional, so the term is functionally meaningless. When you look at the fossil record for any particular species, it is impossible to determine where along that spectrum of evolution a parent species stopped being what they were and began being what the child species ended up being. The child of a chimpanzee is always going to be a chimpanzee, except when you look at several hundred or thousand generations. Then you can see a slow progression from the parent species to the child species, but no one example from the progression can be singled out as a distinct species among its neighbours in the line.

Another problem is that, even if Oliver represented a new species of chimpanzee, what his genetic line would ultimately evolve into would not necessarily or even conceivably be anything like a human, beyond superficial similarities. After all, we do share a common ancestry with all primates, but we could no more spawn a new type of monkey than could a baboon spawn a lemur. It just doesn't work that way.

The argument was settled in 1996 by a geneticist from the University of Chicago, who revealed that Oliver had 48 chromosomes, not 46 like humans, which is consistent with all other chimpanzees. Further study revealed that his facial features were in line with the standard range of variability shown in the Common Chimpanzee, and that his behaviour and habit of walking upright was an aberration brought about because of his exposure to such behaviour in his early life and his lack of chimpanzee companionship throughout his life.

One of the most disturbing things about Oliver's story, is the fact that from 1989 when he was acquired by the Buckshire Corporation – who were responsible for some of the worst mistreatment and inhumane living conditions for animal testing – until his true taxonomy was determined in 1996, it was a widely held belief that he represented a new and distinct evolutionary jump from a wild animal to a more civilized proto-human, yet he was still treated as a side show fixture and a commodity to be exploited. It would be nice to think that humanity has progressed since that point, and that we would treat our companions on this planet with more respect than that, but such ideological thinking is not supported by reality.

Oliver's story is a sad one, he was singled out for being different and held captive for his entire life. He suffered horrible tortures at the hands of scientists (as much as a person researching new eye make-up formulations can be called a scientist) and ultimately died never knowing what it meant to be a chimpanzee. His fate is not unique among his kind, but that makes it no less tragic. He was not a missing link in our evolutionary history, nor his own. He was the product of a biological system that is known for producing oddities and aberrations, but one which we humans seem to have only a rudimentary understanding of thus far. Our pursuit of knowledge should never come at the expense of another living creature.

# The Sounds of History: Acoustic Significance in Ancient Architecture

When we study the ancient world, we have but one sense to use. We can, unfortunately, only view the past with our eyes. As beautiful as the artefacts of our ancestors are, this one dimensional perspective tends to be somewhat restrictive to our understanding. After all, when we consider our contemporary world, we have the benefit of seeing, smelling and hearing all of the various elements that make up that landscape. Not so with the ancient world.

However there are a select few people trying to change that. Those people are working in the field of archaeoacoustics, and though this is a relatively new field of study, great strides are being made in an effort to understand the significance of sound as it pertains to the monuments and rituals of our ancestors. The term archaeoacoustics has been coopted from its earlier use, as it pertained to sounds being recorded in clay pottery and other such objects during their manufacture in ancient societies, so as to be "played back" with the use of modern equipment. This idea was once supported by many in mainstream science, but has recently fallen into disrepute as a result of many failed attempts to verify it through experiment. The term now relates more widely to the study of sound in ancient construction and monuments.

In spite of the fanciful ideas of the more conspiratorial among us, not every ancient monument was constructed to capitalise on resonant frequencies, but some were and they deserve a closer look.

Chanting, a ritualistic form of stylised speech, and the root of all western music, was first used by ancient and prehistoric spiritual leaders in nearly all cultures as a means of furthering or supporting other aspects of ritual. It was meant to bring the participant closer to a religious or spiritual awakening. Chants are used in nearly all religious variants, from modern shamanistic cultures to pagan, Christian, Hindu, Buddhist, and Islamic traditions. It ranges from simple melodies to complex musical structures and depending on the setting, can offer a profound experience to witnesses.

As is common knowledge, sound or music has a profound effect on us humans (and likely on some animals as well). We develop strong associations between musical elements and certain emotions and our moods are often deeply affected by what we hear. For this reason, spiritual or religious chants often have a deep effect on our perception of related experiences. Religious hymns are designed to foster a connection between the congregant and the clergy, and in fact churches the world over are constructed with this in mind. The shape and orientation of the church and its internal elements are painstakingly arranged to optimise the acoustical properties of the space, so as to maximise the effect of song and instrument alike. And this is by no means a new practise.

Nowhere is acoustical significance in ancient construction more striking than in underground temples. There are famous examples of such construction throughout the old world, perhaps the most famous is the King's Chamber in the Great Pyramid of Khufu (or Cheops, whichever you prefer) at Giza in Egypt. Some theorists maintain that the King's Chamber was designed and built to use sound as a resonant booster, to give the Pharaoh a better chance of reaching the afterlife, though this is not a widely held opinion among mainstream archaeologists or Egyptologists. Those same theorists, conspiracy theorists you might say, suggest also that the Hall of Records, an unconfirmed structure or room situated under the Sphinx, has significant acoustical properties as well. This is, for obvious reasons, entirely suppositional of course.

But we needn't resort to conspiratorial fantasy in this case, for there are many ancient monuments and temples that use sound and acoustical properties to their advantage. The underground city complex at Budapest, called the Labyrinth of Buda Castle, which is located under Castle Hill in Buda (which is the west-bank part of Budapest on the Danube river in

Hungary), is said to have special acoustical properties, though since this site is largely a natural formation, it doesn't really count here. It does remain the oldest known example of the shape of a room or cave being used to amplify or resonate sound for ritual purposes.

Other examples, such as the Oracle Room in the Hypogeum of Ħal Seflieni in Paola, Malta (Greece) offer much to study. Hypogeum means 'underground' in Greek, and in this case refers to a subterranean labyrinthine structure of the Seflieni phase of Maltese prehistory (3000-2500BC). It consists of several passages and chambers, of which the Oracle room is the smallest. With its delicately painted ceiling, the Oracle room boasts the most powerful or effectual resonant chamber in the ancient world. Even muted sounds made in this chamber resonate and amplify, which has the effect of distorting the sound and making it seem like it has a divine origin (or that it hadn't been generated by any source in the chamber). Today the hypogeum is a necropolis, containing the remains of some 7000 prehistoric Greeks, but at one time it was used for religious ritual.

Another site, Chavín de Huantar in the Peruvian Andes, is a large city ruin that was built by the pre-Incan culture known as the Chavín in approximately 1200BC, though the area is thought to have been occupied as early as 3000BC. The site has buildings, ruins, temples and other artefacts.

Ancient visitors and priests at Chavín de Huantar would have been privy to an experience not found anywhere else. The buildings were constructed using a highly specialized combination of shafts, corridors and surfaces, all designed to make a series of echo chambers, in which sounds – often conch shell trumpets, called pututus, being blown by priests outside of the structure and chanting, as well as water running in streams under and around the buildings – would seem otherworldly.[1] Add in the psychotropic effect of ritual consumption of San Pedro cactus juice (and possibly other substances, like ayahuasca), and one can easily see how a pilgrimage to such a temple would have been a profound spiritual experience.

Perhaps the first archaeoacoustic researcher, Iegor Reznikoff, an anthropologist of sound with the Université Paris Ouest, found, in the 1980's, that there is a connection between the location of prehistoric artwork in the caves at Lascaux (and other ancient cave sites in southern France, where the oldest known human art is found from 25,000BC) and the acoustic resonance of those same locations.[2] Reznikoff and a colleague mapped such caves, highlighting areas of acoustical significance and found that those areas coincided with areas that held the most works of prehistoric art.[3] Which suggests a defined ritualistic process to the painting, and may have been prevalent among prehistoric artists.

Acoustic resonance is a feature of many natural caves, and it's likely that this natural feature was the primary motivator in the development of acoustics in ritual sites and practices. Modern technology allows archaeologists to identify and study such features of ancient sites, and in most cases the research is inaccessible to the amateur. However, there are branches of this endeavour that are within reach of anyone who can get themselves to the locations in question.

Recently, a team of researchers have been using sound to study the world famous Stonehenge megalithic site in Wiltshire, England. According to experts from London's Royal College of Art, Stonehenge holds more mystery than meets the eye. For many years, enthusiasts and researchers have held that Stonehenge had an audio component, either in its use or construction. Many visitors report that chants and music seem to resonate in a strange way at various points within and around the structure, but new insights seem to suggest that the stones themselves were musical instruments.

Research recently published in the *Journal of Time & Mind*, suggests that the bluestones – the smaller stones that make up the interior of the monument – actually have acoustical properties and may have been selected for that reason. [4] It turns out that the stones resonate in a peculiar way when struck with a hammer or other instrument, and generate a wide range of sounds. Researchers even found what may be evidence of hammer or stone strikes on several of the stones, indicating that they're on the right track.

This research, with the input of other experts, suggests that many of the standing stone sites throughout the UK may have had, as a central feature, an acoustic nature. [5] It may be that Stonehenge and other standing stone circles and like monuments were built as musical instruments, to be used in conjunction with or as a part of ritualistic gatherings and celebrations.

The same may be true for monuments all over the world, as is highlighted by researchers such as Michael Tellinger, who demonstrates in a video on his YouTube channel the acoustic properties of artefacts found at Waterval Boven, South Africa.

There is no denying it, sound has played a central role in the development of not only human spirituality and culture, but also in architecture. While most of our history can only be relayed in terms of visual artefacts and writing, the aural history of our ancestors just begs to be heard. And when you consider the fact that resonant sound has been a significant part of human life for upwards of 27,000 years (at least), it's no wonder so many people feel so passionately about music and its makers.

[1] Brooks, Michael. Was sound the secret weapon of the Andean elites? NewScientist Magazine – September 2008 http://www.newscientist.com/article/mg19926721.700-was-sound-the-secret-weapon-of-the-andean-elites.html?page=1

[2] Starr, Douglas. Notes from Earth: Echoes from the Distant Past. Discover Magazine – November 2012 http://discovermagazine.com/2012/nov/03-echoes-from-the-distant-past#.UsCjmvRDsid

[3] American Institute of Physics. "Music Went With Cave Art In Prehistoric Caves." ScienceDaily, 5 Jul. 2008. Web. 29 Dec. 2013.

http://www.sciencedaily.com/releases/2008/07/080704130439
.htm

[4] Paul Devereux, Jon Wozencroft. Stone Age Eyes and Ears: A
Visual and Acoustic Pilot Study of Carn Menyn, Environs, Preseli,
Wales. Time & Mind
http://www.tandfonline.com/doi/full/10.1080/1751696X.2013.
860278#.UsCuvfRDsie

[5] Sarah Griffiths, Amanda Williams. Stonehenge 'was a
prehistoric center for rock music': Stones sound like bells,
drums and gongs when played. DailyMailUK December2013
http://www.dailymail.co.uk/sciencetech/article-2515159/Why-
Stonehenge-prehistoric-centre-rock-music-Stones-sound-like-
bells-drums-gongs-played.html

# Teleportation: From Ancient Myth to Modern Science

Being a die-hard fan of Star Trek, I basically grew up accepting the idea that people could be beamed from one location to the next. They made it look so easy; you just stepped onto the lighted pad while some guy in a red (or yellow) shirt hit a few icons on his control board and after a few wibbly lines and sparkles, away you went. They were never really clear on exactly how it worked or how far they could send you, but it must have been anywhere from a few hundred thousand miles to a million. What a way to travel!

Of course, that's a TV show. A particularly good TV show in my opinion, but a fictional construct nonetheless. Mr. Roddenberry was faced with a conundrum when he created a show based on interstellar travel, including visits to all manner of alien worlds. How do we get our characters from the ship to the surface without endless voyages in shuttlecraft or what have you? Easy, we invent a machine that magically transports them in an instant! But did Roddenberry really invent the idea? Well, no, he didn't.

The idea that a person or thing can be magically transported from one location to another is actually quite an old one. It has shamanistic origins, and there are accounts, arguably, in the Bible, but it likely predates the Biblical period. Those Biblical accounts, Ezekiel 11:1, and in the story of Daniel and the Lion's Den from the Hebrew Bible, tell of the mystical phenomenon of bilocation, where a person is observed in two places at once, often impossibly far apart. This idea is also found in Vedic traditions, Buddhism and many other spiritual customs. The story from the Holy Quran, of the Prophet Muhammad's Night Journey from Mecca to Jerusalem, is sometimes thought of as another example.

The idea has a few names too: bilocation (also given as bi-location), apportation (or to aport), teletransportation, or more commonly, teleportation. These terms all have slightly different meanings, but all refer to the same phenomenon. The term teleportation was first coined by the inimitable father of paranormal research, Mr. Charles Fort in 1931, in his second non-fiction book titled Lo!.[1] In it he described various events and happenings revolving around the idea and presented his thesis that, by way of a "cosmic joker", certain objects and people could be transported over great distances by unknown means. Fort connected many disparate phenomenon with teleportation, from telekinetic apportation, which is associated with spiritualistic séances and mediums, to missing persons cases and even weird rain (strange items and/or animals falling like rain, often from clear skies). *"Mostly in this book I shall specialize upon indications that there exists a transportory force that I shall call Teleportation."*

But as mentioned, the idea long predates Fort and the spiritualism movement of the late 19th century. The problem, as with any Fortean subject, is that the older the account, the less credible the source. There are many stories from almost every culture that feature an event resembling Fort's idea of teleportation, but it's exceedingly difficult to pin down details, and thus we are forced to look at them as apocryphal myths. Of course, the more modern accounts don't really offer that much reliable information either.

Apportation gets a bad rap, resulting from the questionable methods of mid to late 19th century and early 20th century mediums and spiritualists, who used sleight of hand and outright trickery to dupe sitters into believing objects, such as flowers, stones, perfumes, and small animals, were either spontaneously disappearing or appearing (or both) during a séance. Almost every account from this period has either been debunked or is considered to have been hoaxed, but there are a few worth mentioning. The amazing story of the Pansini Brothers is one such account.

The Pansini Brothers, the sons of Signor Mauro Pansini, an Italian building contractor, were considered to be "mediumistic children". Following what was said to have been poltergeist activity in the family's older home in 1904 and ongoing accounts of the older son speaking in tongues, the boys, Alfredo (10) and Paulo (8), we mysteriously transported a distance of ten to fifteen miles from the home in mere minutes. Apparently there were multiple teleport events involving both boys, and on one occasion, in the presence of a bishop Bitonto, the boys vanished from the room as their mother and the bishop discussed means for ending this "obsession".[2]

Despite fairly close scrutiny by Italian scientists at the time, no explanation was ever found for the events.

Another notable account of teleportation is that of Damodar Ketkar of Poona, India. Ketkar, described as a young child in the grips of a "poltergeist persecution", suffered a teleportation event on April 23, 1928. According to a letter written by the boy's British Governess, Miss H. Kohn, Damodar materialised in front of her and said to her "I have just come from Karjat!" (Which is approximately 63 miles from Poona)

Kohn noted, with some enthusiasm, that the boy's posture upon materialising was "...of a person who has been gripped round the waist and carried, and therefore makes no effort but is gently dropped at his destination."[3] He apparently suffered no ill effects from the experience.

This case is unique and particularly interesting, as it's the only known case of a person's teleportation arrival being witnessed independently. As with the others though, this tale stands, and will remain, uncorroborated.

Of course, anyone who stays abreast of modern technological advancements, is aware that scientists are working on making the Star Trek transporter a reality. This research is in the realm of quantum physics, and it involves what Einstein called "spooky action at a distance", otherwise known as quantum entanglement. A certain level of success has been achieved in the field of quantum teleportation, but we're still far from zipping through space, from planet to planet, for various complicated reasons.

It is reasonable to think, though, that in time our greatest scientific minds will master the science and bring us something like a sci-fi transporter, but as Eric W. Davis concluded in his 2004 special report to the US Air Force Research Laboratory on teleportation physics: *"At present, none of the theoretical concepts explored...have been brought to a level of technical maturity, where it becomes meaningful..."* [4]

[1] Charles Fort. Lo!. Claud Kendall (Publisher) 1931 New York [Online annotated version]: http://www.resologist.net/loei.htm

[2] Lapponi, Joseph. Hypnotism and Spiritualism. New York: Long-Mans, Green and Co. 1907

[3] Price, Harry. An Indian Poltergeist with Miss H. Kohn. Psychic Research (New York) March 1930

[4] Davis, Eric. W. Teleportation Physics Report. Air Force Research Laboratory, Air Force Materiel Command – August 2004 AFRL-PR-ED-TR-2003-0034 http://www.rense.com/1.imagesG/teleport.pdf

# Dreams & Near-Death Experiences: Two Sides of a Different Coin

Where do the images in your dreams come from? Are they the product of pure imagination, or do they have a more down to earth origin?  This might seem to you, to be a fairly simple question, but rest assured, it's anything but.

There are in excess of eleven scientific theories on dreams, from both a psychological and a neurobiological perspective.  There is the Freudian view of dreams and the Jungian view of dreams.  Dreams have a cultural meaning in terms of ancient history, philosophy, theology, literature and pop-culture.  Some believe they can be prophetic, others think they're the manifestation of an alternate reality.  To some they are a nightly escape into a world of fantasy, for others they are frequently a horrific adventure into past hurts and fears.

Whatever they mean, which ultimately is a highly subjective and semantic idea, there are some facts about dreams of which you might not have been aware.

There is, without question, a defined physiology to dreams, a physical process undertaken by the brain, the mechanics of which are relatively well understood.  It is our subconscious mind playing movies for our benefit while we sleep, right?

Well no, not actually. Dreams occur most often during REM sleep.  That is, the sleep cycle characterised by Rapid Eye Movement.  The average person spends, or will spend, approximately six years of their life dreaming, in nightly bouts lasting between five and twenty minutes.  There doesn't appear to be a single area of the brain responsible for generating dreams, but little is known about their precise origin, despite continuous testing and research for centuries.

So, where do the landscapes and characters of our dreams come from?  Without delving into a discussion of the merits of Freudian or Jungian archetypes, which are more interesting in discussion than in practise, there are a few theories that shed some light on the subject.

Since 1976, when J. Allen Hobson and Robert McCarley turned Freud's theory of unconscious wishes on its head, most researchers have yielded to the idea that dreams are, in some fashion, the result of our subconscious mind sorting out short- and long-term memories from our waking lives.

Along a progression in thinking, from the Activation Synthesis Theory, which speaks more to the neurobiological origin of dreams, to the less well defined theories of long-term memory excitation and the strengthening of semantic memories, it seems clear that memory plays a key role in the dreamscape.

Published in the journal Science (October 13, 2000 issue), are the results of the inspired research of the Associate Professor of Psychiatry at Harvard Medical School, Robert Stickgold PhD. Stickgold is, among other things, the Director of the Center for Sleep and Cognition.

His paper, titled *Replaying the game: hypnagogic images in normals and amnesics* tells the story of a group of people playing video games, of all things.

Stickgold and his colleagues used the video game Tetris to study the function of memory in dreams, and they found, as may not be surprising, that when people played the video game for a set period of time prior to sleeping, their dreams featured elements from the game. From geometric shapes and landscapes, to activities such as sorting objects in the dream. This may not seem particularly interesting, but consider that he tried the same experiment with amnesiacs, or people who suffer from amnesia, a neurological condition that prevents sufferers from converting short-term or semantic memory into long-term memory. There are different forms of amnesia, ranging from the Hollywood style wiping of a person's long-term memory and identity to other types of memory related maladies as noted above.

When Stickgold conducted his experiment on the amnesiacs he found, quite remarkably, that they too dreamed of geometrically themed landscapes and activities following a set duration of game play. [1]

Interestingly, this seems to prove, with room for discussion, that dreams are in fact the process of the subconscious mind cementing learned information into either semantic memory or episodic memory – which is related to factual knowledge, i.e. 2+2=4, rather than subjective experiences. Further research by Stickgold and others has further confirmed this idea in recent years.

This means that the imagery you see or experience while dreaming isn't the product of pure imagination, it comes directly from memory. The faces and buildings and other sensory aspects of your dreams are taken directly from, or are amalgamations from recent memories. When you dream of people or places that you don't readily recognise it may be because your mind has twisted the details ever so slightly in its attempt to make sense of the information, or that the memory involved was insignificant to your waking consciousness and therefore isn't something you readily recall, or possibly a combination of both processes.

This is all very interesting, especially if you're a student of psychology, or are particularly fond of Freud, but it has an impact on something you may not immediately realize.

As reported, Dr. Sam Parnia, Critical Care Physician and Director of Resuscitation Research at Stony Brook University School of Medicine in New York, has, through his research in conjunction with the AWARE Project (The Nour Foundation – Human Consciousness Project) likened the phenomenon of Near-Death Experience to that of a dream state.[2] *"We've certainly found in our studies … that if we manage to get to patients immediately after waking up — which is not easy at times — and talk to them, they tend to remember more, and if you go back and reinterview them within a couple of days, they tend to have forgotten their experiences, possibly. So we think that probably many more people have these experiences — if perhaps not even everyone — but somehow their memories get wiped in the same way that most of us — if not all of us —*

*dream every night, but somehow there's a disruption to the memory circuits that allow us to recall our dreams the following day."*[3]

This seems a reasonable comparison, NDE's, which often are reported to have Out-Of-Body Experiences associated with them, are described in much the same way as dreams. Though one thing is quite different about NDE's, in roughly 80% of reported NDE's the imagery is universal or archetypal among those who experience it. Meaning, in simple terms, that those who have Near-Death Experiences often report strikingly similar landscapes, events and characters in the dreamlike world of the experience.

This has been held up as strong evidence that NDE'ers are in fact experiencing a real event. That they are actually meeting with loved ones and with religious characters and are travelling in a real, albeit non corporeal place to whatever end, and are ultimately being pulled back from that place or journey upon resuscitation.

This has potentially far reaching implications, such as bringing us closer to answering the mind-body question. That is, do we have a soul? Or, is there an afterlife?

Those questions are without a doubt very much unanswered as it stands, and while Parnia's research pushes forward, though admittedly not with the goal of answering those questions directly, it seems we may find ourselves closer to an answer in the near future.

The above though, might give you the wrong impression. NDE's are not like dreams, as convenient as Parnia's analogy above may be.

Dr. Kenneth Ring, Emeritus Professor of Psychology at the University of Connecticut and one of the leading researchers in Near-Death studies, has outlined in his 1999 book Mindsight: Near-Death and Out-of-Body Experiences in the Blind that, even without the benefit of sight, blind people share in the common archetypal imagery known to sighted, Near-Death experiencers.[4]

This sort of flies in the face of the idea that NDE's are in any way similar to dreams, other than superficially. One of Ring's research subjects, Vicki Umipeg, a forty-five year old blind woman, remarked about her experience: *"This was the only time I could ever relate to seeing and to what light was, because I experienced it."*[5]

Vicki was blind from birth, and as researchers have known for decades, if not centuries, blind people do not dream in visual terms. They experience dreams in terms of other sensory perceptions, such as touch, smell and sound etc. If a person lost their sight at some point later in life, they can experience dreams that incorporate limited visual stimulus, relative to the length of time they've been blind. In simpler terms, blind people don't have visual dreams, because they have no visual memories.

This supports the notion that dreams are the product of memories, as demonstrated by Stickgold et al. Though it quite thoroughly dismisses the idea that NDE's are dreamlike. If they are not similar to or related to dreams, and the imagery experienced during an NDE are not the product of memory as dreams are, what does that say about where the imagery of NDE's comes from?

Many modern theories coming out of neurophysiology and psychology have suggested, with varying success, that the environment and characters reported with NDE's are the result of some unknown neurological or neurodegenerative process associated with the early stages of death. Whether the instant decay of neurons and synapses or de-oxygenation of blood in the brain, or even changes in the quantum state of the brain being perceived by failing synaptic functions, the problem is that the specific archetypes reported seem to be generated independently of the memories of the patient.

If the rich and often alien environments reported with NDE's are the product of imagination, this is counterintuitive when considered alongside the above theories. Can a brain with no electrical activity and quickly degenerating physiology be expected to generate a vivid and hyper-realistic dreamscape, the likes of which could scarcely be replicated in actual dreams?

This, of course, is entirely suppositional, and doesn't speak to any of the other issues tangled in amongst the mind-body question, but it does seem to give much food for thought. NDE's and their associated OBE's seem not to be associated with the memories of their experiencers, and as such, their imagery need to be accounted for in some other way. Whether that's through neurophysiology or quantum mechanics is yet to be discovered, but we do seem to be tantalisingly close to an answer nonetheless.

[1] Stickgold R, A Malia, et al. *Replaying the game: hypnagogic images in normals and amnesics*. Science. 2000 Oct 13; 290(5490):350-3. [PMID: 11030656]
http://www.sciencemag.org/content/290/5490/247.1.summary

[2] Nour Foundation – Human Consciousness Project:http://www.nourfoundation.com/events/Beyond-the-Mind-Body-Problem/The-Human-Consciousness-Project/the-AWARE-study.html#sthash.R1si2gMb.dpuf

[3] NPR.org – 'Erasing Death' Explores The Science of Resuscitation. February 20, 2013.
http://www.npr.org/2013/02/21/172495667/resuscitation-experiences-and-erasing-death

[4] Ring, Kenneth. Mindsight: Near-Death and Out-of-Body Experiences in the Blind. iUniverse Books, ISBN-10: 0595434975

[5] Williams, Kevin. People Born Blind Can See During A NDE – Dr. Kenneth Ring's NDE Research of the Blind:
http://web.archive.org/web/20080117054508/http://www.near-death.com/experiences/evidence03.html

# What Does The Soul Weigh? Kind of a Heavy Subject!

In the aftermath of the flurry of articles I've read recently, which point out the problems we researchers have with the vast mountains of information available through the internet, it seems particularly apt that I should happen upon the conspiratorial and incredible story of the weight of the human soul today.

Regular readers are aware of my growing obsession with the mind-body question. That is, do humans have souls? A question that first reared its head in ancient Greece, with the great philosophising of Heraclitus (c. 475 BCE). Despite the passionate assertions of a great many people, this question remains unanswered.

Outside of the sometimes highly satisfying philosophical ideas associated with this question, it seems the only way to answer this question with any certainty is through scientific investigation. Much of that has taken place in the last few decades, from Penrose & Hammeroff's Orch-OR theory of the quantum soul, to Ervin Laszlo's Akashic Field Theory, to Parnia's research through the AWARE Project, there's no shortage of ideas to read about.

One particularly intriguing feather in the cap of those who claim success in this area of study is the work that's been done to weigh the human soul. This idea featured prominently in Dan Brown's book The Lost Symbol, which I quite enjoyed even though it's not his best work. In the book, Brown described work that had been done by the Institute of Noetic Sciences, though in the book he gives all the credit to the female protagonist working alone and on behalf of the Smithsonian Institute's top secret research division. He describes a highly scientific and technologically advanced apparatus used to dynamically measure the weight of people as they died. He gave no detail regarding the results, however.

As everyone knows, fiction is fiction, and Dan Brown is famous for weaving what appears to be truth into his stories, ultimately fooling a great number of people into believing it's all based on fact.  In this case, as with others, it was not.

I found this concept, that is, that the soul could be weighed, to be of great interest to me personally and so I looked into it.  It turns out there is a basis in truth here...sort of.

In 1901 Dr. Duncan McDougall, a physician from Haverhill, Massachusetts, undertook an inspired experiment to determine how much the soul weighed by measuring the body-weight of 6 patients prior to and following death.  He found, as the story goes, that the soul weighs 21 grams.  This result is an averaging of the body-weight difference between patients from a few moments after death.

His experiments, which he also conducted on dogs and apparently found an agreement between species, were eventually discovered and reported through the Journal of the American Society for Psychical Research, and the journal American Medicine, as well as The New York Times.

The problem is, that there were problems.  His methodology was so sloppy that no one could replicate his results.  And, since this was the turn of the 20th century we're talking about, the available technology was less than reliable.  It stands though, that McDougall tried, and had limited success, in exposing that there seems to be a difference between the weight of a body before death and after.  It's an easy jump from there to believing the mind-body question answered, but it's not.  Most of mainstream science regards his conclusions as false, or simply wrong. [1]

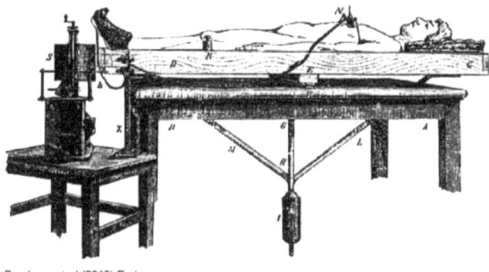

Sandrone et al (2013) Brain

This wasn't the first such attempt either. Early Italian neuroscientist Dr. Angelo Mosso conducted a similar series of experiments in approximately 1884, with his 'metal cradle' or 'machine to weigh the soul'. Rather than measuring the difference between alive and dead weight, he believed he could measure an increase in the weight of the head of a subject, during cognitive effort. His results were less than impressive, for various reasons. [2][3]

McDougall's and Mosso's experiments were not, however, what Dan Brown was talking about. He most likely was referring to a German study conducted in 1988 by two scientists named Becker Mertens and Elke Fisher. In their study, Mertens and Fisher weighed some 200 terminally ill patients and found, universally, a difference of 1/3000th of an ounce between life and death. It seems the soul weighs roughly 0.01 grams. Their results were published in the German science magazine Horizons, and these results are oft cited and held out as proof that the soul exists.

Now, there are some methodological problems here too; namely that air leaving the patients lungs could account for the weight difference, or some instantaneous decay event, possibly releasing gas held inside the patient's cells. These and other criticisms have been levelled at this and at McDougall's results, but there's an even bigger problem at play.

The whole thing is a hoax, Becker Mertens and Elke Fisher do not exist, nor does the magazine Horizons.[4] No such research has been undertaken and the so-called evidence is entirely fabricated.[5] *"This type of miss information (sic) is a growing problem, especially for people overly reliant on the web for information. Such irresponsible fabrication does not serve the scientific community or the general public."*[6]

The above hoax, as I would dare to call it, has been retold and blogged about many times, as though the whole things is completely true. Most notably by new age magazine New Dawn (special issue 15, page 70) and in a Weekly World News article.

As was highlighted perfectly in his New York Times article of October 25, 2013, Steven Schlozman M.D., warns of how easily things like this can get out of control, and how damaging they can be to not only our understanding of the issues involved, but also to our cultural and social evolution. In his case, the culprit was harmless joke, in this case it may not have been meant in such a pithy tone. With works like Brown's Lost Symbol clouding the issue even further, is it any wonder the layman, the regular Joe (or Josephine) has trouble sorting out fact from fiction?

[1] Mikkelson, Barbara; Mikkelson, David P. (October 27, 2003). "Soul Man"– Snopes

[2] Sandrone S, Bacigaluppi M, Galloni MR, Cappa SF, Moro A, Catani M, Filippi M, Monti MM, Perani D, & Martino G (2013). Weighing brain activity with the balance: Angelo Mosso's original manuscripts come to light. Brain PMID: 23687118

[3] Neuroskeptic – A Machine to Weigh The Soul: http://blogs.discovermagazine.com/neuroskeptic/2013/05/21/a-machine-to-weigh-the-soul/#.UmwiW_kWLCs

[4] The Tribal Scientist – New Horizons, old hoaxes: http://tribalscientist.wordpress.com/2012/03/21/new-horizons-old-hoaxes/

[5] Kennedy, Chad, PhD. Spiritual Evolution: How Science Redefines Our Existence. (Authorhouse 2011) ISBN-10: 1467024147. Pg. 166.

[6] Kennedy, Chad, PhD. Spiritual Evolution: How Science Redefines Our Existence. (Authorhouse 2011) ISBN-10: 1467024147. Pg. 166.

## You Can't Change the Past...Says Who?

Everything you experience is in the past. In fact, the nature of reality dictates that everything you will ever experience, everything you can experience will have already happened by the time you experience it. Confused? Allow me to explain.

Light travels fast, as you may already know. Really fast. Mind bogglingly fast. 299,792,458 metres per second to be precise. Or in more relative terms, roughly 300,000 kilometers per second. That's pretty much the ultimate speed limit of the universe. According to Albert Einstein's Theory of Special Relativity, nothing can travel faster than light, but if you could, which you can't, but if you could, you'd be able to plant yourself at any point along the space-time continuum. That is to say that you could travel to any place or time that you choose, with certain caveats, the most relevant being that you can't. Still confused? I'll continue then.

When you look to the stars at night, those twinkling little dots of light in the sky, what you are seeing is the light emanated by distant suns. Some with planets around them, others without. That light that you see is confined by the same speed limit as the light our own sun emits. That means that a distant star, one perhaps several thousand light-years away, isn't exactly as it appears in the sky.

When the light leaves its parent star, it's travelling at 300,000 kilometers per second, but there are many millions, or even billions of miles between stars. For example, the nearest star to our solar system (besides our sun) is called Proxima Centauri and it is approximately 4.24 light-years from Earth. That means that it takes light, the light emitted from the star, approximately 4.2 years to reach Earth, and subsequently...your eye. That is precisely where the term light-year comes from, it is the standard by which galactic distance is measured, and all because light travels at a constant speed.

So, in case you slept through high school science class, that means that all you can ever see of the stars in the night sky, is their past. Their distant past in most cases. In some cases we're talking about billions of years distant.

You may be thinking that this is all well and good for stars that are thousands, millions or even billions of light-years away, but how does this mean that everything I experience is in the past?

Aha! This is where it gets interesting.

Like I said above, light speed is constant. It's dependable, like a trusted neighbour, or your mom. It will always travel at 300,000 kilometers per second. Well, that's not exactly accurate, but outside of some very specific circumstances it is constant. That means that the light emitted from the light bulb in your bedroom also travels at that speed. Now, unless you're mega-rich and have a mansion the size of a galaxy, that light probably only has a few feet to travel, so in any perceptible way it happens instantaneously. But our perceptions aren't really trustworthy.

Even over a distance of a few feet it still takes time for the light to travel from the bulb to the wall, where the wall's surface, depending on its colour, absorbs all but a very narrow band of the spectrum, and the light is then reflected away from the wall and into your eye. This is where it impacts the special cone shaped cells that line your retina, and you see it. This all happens in a very short fraction of a second, depending on the size of the room, but rest assured, it happens constantly, everywhere...in every room with a light bulb that is turned on.

So you see, that elapsed time between the light streaming out of the light bulb, hitting the wall and being reflected into your eye takes time. It means that the state of the wall, or whatever object you're looking at, could conceivably be different than it was when the light hit it, than it appeared when that light hit your retina. Not convinced? OK well, there's more.

The speed limit of light is not the only factor that contributes to the fact that everything you experience is in the past. After all, that's only one of our five senses (as though we have only five). In fact, it's our senses that really prove this concept to be true.

Let's stick with light for the time being. After that light jumps through all the hoops I laid out above, it does indeed strike the back of your eye. It's collected by those little cone shaped cells on your retina (there are also rod shaped cells, which do the same thing, in a slightly different way, but this isn't important to this discussion). Those little cones do a little magic (not really) as they convert the different wavelengths of light and the varying brightness of that light into a complex series of electrical impulses that travel up the optic nerve, through the back of the eye and into the visual cortex of the brain. There, those electrical impulses are sorted and correlated with a whole library of memories and the entire jumbled mess is converted into an image. Not like a jpeg or something. Well, actually, it's a little like a digital image, but that's another story.

Next, your visual cortex serves up a conceptual model of that image to your conscious mind, which you experience as visual information that has been encoded with spatial information and made relevant to your world view through a comparison to memories, dreams and previously learned information. It's a complicated process, though somewhat elegant if you think about it, and it happens constantly. Even when your eyes are closed and you're asleep.

I know, I know...I still haven't explained how this means everything you experience is in the past. OK, here it is. That process, like the speed of light itself, happens almost instantly...almost. It does take a certain amount of time for those little electrical impulses to travel up the optic nerve, or the auditory nerve, or the olfactory nerve, or...well you get my point. It also takes a small amount of time for the synapses in your visual cortex, or your olfactory bulb, to collate and cross reference the information against your previous experience, and then finally it takes a further fraction of a second for that information to be served up to your consciousness as a perceptual sense.  The point is...it all takes time.  Not much time, but as with the consequence of the speed of light, the information being received by your eye is technically out-of-date by the time it reaches your consciousness.

The world at large is constantly changing, from nano-second to nano-second and our perception of this world is, in fact, nothing more than a conceptual model served up to our consciousness by the hardware of our nervous system.  Just as you can never technically consider yourself truly in the moment, this fact also proves that your senses are really only giving you a general approximation of your external reality.  You aren't ever actually experiencing the real thing...and what you experience is always filtered and coloured by your previous experience and knowledge.

When you consider the elegant but convoluted means by which your optic system feeds you information, consider also the many ways that this system can break down.  Most of those ways will result in blindness, or partial blindness, but some will just skew the information your brain is being fed.  Which will in turn skew your perception of reality.

All in all, this has been a somewhat wordy way to illustrate that your senses can't be trusted.  This applies to the paranormal world in a very specific way, and I know you got the point.  Try to remember it.

# The Backster Effect: Are Plants Aware?

I would venture to say that every adult who might happen across this post has had some experience with lawn mowing. That is to say that most everyone has observed or participated in the act of grass cutting; making it a common practise and experience in most cultures (though I admit this may be an example of 1st world chauvinism).

In any event, none (or at least very few) of those people would ever give a second thought to the potential pain and brutal destruction going on below their feet. A grand act of herbicidal horror, genocide if you will, an act so heinous that, were it perpetrated against a conscious living being the culprit would be subject to the most capital of punishment and societal scorn. But cutting the grass isn't a crime, it isn't murder (or aggravated assault?), and anyone who suggests otherwise should be facing the condescension of a psychiatric professional...or should they?

Well, I'll tell you, there are quite a few people who might say that grass, or any plant for that matter is a feeling, emotional and communicative entity. Known as plant perception or biocommunication, the idea that plants are sentient living beings is gaining support among the metaphysical / paranormal community. And if this sounds strange to you, consider the common notion that plants respond to the human voice, and are nurtured by a daily dialogue. I'm willing to bet there are many of you that talk to your plants on a regular basis.

The idea that plants respond to human interactions was first postulated by Dr. Gustav Theodor Fechner in 1848. Fechner, an experimental psychologist, suggested that plants are capable of emotion and that one could promote healthy growth with talk, attention and affection. [1] This idea remained largely untested until 1900, when Indian scientist Jagadish Chandra Bose found, through experimentation, that all plants have a form of a nervous system and will respond to electric shock by spasm.

Support for this idea took some time to get off the ground though, until Cleve Backster, a CIA interrogation expert of all things, conducted experiments using a polygraph machine to measure bioelectrical responses in many types of plants. Specifically, in February of 1966, Backster measured electrical resistance in the plant's leaves which would change when the plant was watered.  He found that plants *"...show a pattern typical of the response you get when you subject a human to emotional stimulation of short duration"*.

As a result of his experiments, the idea that plants have a conscious awareness and capability to communicate with the world around them is now known as the Backster Effect. Several books have been published expounding on the Backster Effect and expanding on the original hypothesis, such as The Secret Life of Plants by Tomkins and Bird.[2]  The effect was even tested by the inimitable Mysthbusters (Season 2, Episode 5, 2006), wherein they examined plants using a galvanometer (the main component of a polygraph machine).  They subjected plants to several real and imagined dangers, and even potential psychic influence, though they came up empty, finding no statistically relevant results.

It should be known that mainstream science largely views biocommunication and the Backster Effect to be pure pseudoscience (with some exception, which we'll get to in a minute), dismissing the entire idea because of the fact that plants have no apparent nervous system.  But Hollywood has no such compunctions.  The popular M. Night Shyamalan movie The Happening, staring Mark Wahlberg and Zooey Deschanel illustrated the dark side of the Backster Effect quite well.

Wahlberg's dismal acting aside, is it possible that plants, the plants that surround us, from blades of grass to massive haunting forests, are actually plotting our collective demise? After all, we do pretty much everything we can to annoy, afflict and assault them on a daily basis.  If they have emotions and the ability to express them, aren't we due a retaliatory slap in the face about now?

As it turns out, Shyamalan and Wahlberg weren't far off the mark. In a paper published in the scientific journal Ecology Letters on May 9th 2013, several scientists have found that plants actually do communicate, with each other if not the rest of the world. [3][4] The paper, authored by a team of scientists from the University of Aberdeen, Scotland, the James Hutton Institute and Rothamstead Research, details how plants – they used broad bean plants – use an underground network of fungus called mycorrhizae to signal other nearby plants of impending danger.

In their study, they subjected the plants to simulated attacks by aphids (small insects that feed on many different plants), and they found that through this fungal network, the plants were able to warn the other plants of the attack, giving them time to deploy defensive strategies such as releasing chemical pheromones to discourage the aphids and to attract wasps, the aphid's natural predators.

Of course, there are several plant species that are known for their aggressive predatory adaptations, such as the Pitcher Plant and the Venus Fly Trap, but mainstream science tends to ignore these examples of plants that actively interact with their environment. What if all plants are ultimately capable of affecting their environment in more subtle and potentially dangerous ways?

So the next time you're mindlessly running the lawnmower over the front lawn, or when you're mercilessly yanking weeds from your garden, think twice and be prepared to defend yourself against the uprising of the dandelions!

I know many people feel a deep connection to plant life and those people usually aren't shy about describing their relationships with their garden's inhabitants. If you're one of them, I invite you to share your story in the comment section below.

[1] Michael Heidelberger Nature from within: Gustav Theodor Fechner and his psychophysical worldview 2004, ISBN 0-822970775, p. 54

[2] Peter Tomkins & Christopher Bird, The Secret Life of Plants (1973) ISBN 0-06-091587-0

[3] David Ferguson (TheRawStory.com), Plants Use Underground Fungus Network to Send "Distress Signal" to Each Other, http://www.rawstory.com/rs/2013/05/10/plants-use-underground-fungus-network-send-distress-signals-to-each-other/

[4] Ecology Letters (2013): http://onlinelibrary.wiley.com/doi/10.1111/ele.12115/abstract

# Webbot: The Prophet of the Future

Why do all famed prophets, such as Nostradamus or Edgar Casey, always report fire and brimstone type predictions? Why is there never a prophecy that mankind will have a day of peace and illumination, on, oh I don't know…a Tuesday? (Yes, I am aware that some prophecies have been positive in nature, but the vast majority are negative)

Perhaps the reason is simply the nature of humanity, and I don't necessarily mean that we're destined to destroy the planet and ourselves along with it, which we very well might be. I mean that perhaps the reason is that all of our prophets are human, with fears and flaws and crushed personal aspirations. Or are they? Well people…I give you Webbot!

Ok, so maybe Webbot isn't really all that new, but it is, well, novel. I've written about prophets before, and perhaps I wasn't exactly charitable when I did so. I mean, when you have to change whole words and phrases in a prediction to prove, after-the-fact, that the prediction actually did predict an event – such as must be done with a large number of Nostradamus' quatrains- then you've opened yourself up for some criticism.

Webbot, however, has been 100% accurate in all of its predictions! Well no, actually it hasn't. But shouldn't an electronic, internet-wide, uber-surveillance wonder-algorithm using super bot be able to predict humanity's future with 100% accuracy? That depends on who you ask.

Developed in the late 1990's by Clif High and his associate George Ure to predict stock market trends, The Webbot Project, as it is now called, operates through the website www.halfpasthuman.com[1], where High sells his predictions piecemeal. High claims that his spider-like web robot can accurately predict future events by crawling the web and analysing "web chatter" to identify trends in global human emotion. He treats his algorithm as top secret and is rather tight lipped about how The Webbot Project works. High also has a YouTube channel: The Webbot Project, where he expounds on a wide array of subjects relating to metaphysics, prophecies and the paranormal.

The Webbot Project has been analysing the internet for more than two decades and in that time it has had its share of hits and misses. Much of the predictions you'll find via the website above are rather mundane and cryptic, but some of them have been more sensational, such as:

- The 9/11 Terror attack
- The 2003 Northeast Blackout
- The 2004 Indian Ocean Earthquake and subsequent Tsunami
- Hurricane Katrina in 2005

Now, the legitimacy of these predictions is in question and they are widely regarded as postdictions – predictions that can only be attributed to events after the event has happened, often with liberal editing of the original prediction.

The Webbot Project does, however, have a long list of misses, such as a massive earthquake in Vancouver, Canada and the Pacific Northwest was predicted to occur on December 12, 2008. The US dollar was predicted to completely collapse, or Israel was to bomb Iran in 2011 and in reaction to this crisis, the administration of U.S. President Barack Obama was to be thrown into major chaos ten days later. And most notably Webbot predicted a major catastrophe in 2012, relating to the Mayan Prophecy and a possible magnetic pole shift for the planet.

These failures, especially the 2012 catastrophe prediction have brought The Webbot Project into disrepute, but much like its human counterparts in history, none of this has done anything to dissuade true believers. High continues to churn out prediction after prediction even though those efforts have been criticised citing the ambiguity and gloom and doom nature of the predictions as major faults. Or as Tom Chivers of The Daily Telegraph says: *"...the internet might plausibly reveal group knowledge about the stock market or, conceivably, terror attacks [but] it would be no more capable of predicting a natural disaster than would a Google search, ... the predictions are so vague as to be meaningless, [and] the prophecies become self-distorting."*[2]

Ultimately, is Webbot just as fallible and biased as its human contemporaries? I would say so. While it may be a technological marvel, cold, calculated and robotic, it is fed by human emotion, using our digital voices as the basis for its predictions, and as long as there is a human element to the process, our biases, fears and dreams will be a major part of its prophecy.

What do you think about The Web Bot Project? Should we trust to an artificial intelligence in seeking answers about what the future may bring? Voice your opinion in the comment section below.

[1] See whois.net data for halfpasthuman.com: http://whois.net/whois/halfpasthuman.com

[2] Chivers, Tom (24 September 2009). "'Web-bot project' makes prophecy of 2012 apocalypse". The Daily Telegraph (London).

# History and Literature Stuff

## Hard to Kill: Rasputin and Other Immortals

The human body is a pretty fragile thing, when you think about it.  Maybe not as fragile as movie makers try to make it seem, but still far less durable than a lot of other life forms on this planet.

In the movies they'd have you believe that, at least for nameless bad guys, you can be killed instantly with a good swift kick to the shin.  But if you're a hero (or anti-hero), you can brush off even nuclear blasts with little more than an endearing scratch or two.

Real life, of course, isn't like that.  Introducing foreign bodies into your own, especially of the pointy variety, is decidedly unadvised.  It's generally a good idea to avoid the business end of guns, and most will tell you that knives do not belong in your belly (or your back, but that's another article).

The average person holds about five litres of blood, or one and a half gallons.  It's roughly 7% of your body weight, so the bigger you are, the more red stuff you carry.  Of course, if you spring a leak you have a problem.  That same average person can, in theory, easily survive a loss of blood ranging from 10-15% of capacity (or about half a litre), but beyond that, things get dicey.

Contrary to popular belief though, having a bullet rip through your innards isn't strictly fatal.  It's actually the blood loss that results from all the holes that will kill you.  Yes, there are some other things going on when you've been perforated, but as the spunky female supporting character will always tell you, you've got to stop the bleeding.

Though, you realise that this is all based on a best case scenario, right?  Well, maybe not best case, you've just been shot, after all.  The point is, there are lots of people who defy the odds of survival; people who not only laugh in the face of death, they pretty much give it the full-Monty.

Perhaps the most famous and bewildering tale of such defiant survival is that of Grigori Yefimovich Rasputin. His story is widely misrepresented, however.

Rasputin, as I'm sure you're familiar, was something of a sensation in Russia at the turn of the 20th century. A most controversial figure, he was said to be one of the most powerful faith healers ever to have lived, he was (depending on who you ask) an occultist extraordinaire, and was certainly a most influential spiritual guru of the time. And above all, he was a trusted advisor to the Tsars, and, according to historians, was the primary catalyst for the fall of the Russian Monarchy.

Whole volumes have been written on who Rasputin was, how he lived, and how he died. Strange as he was, and despite what certain films would have you believe, he was definitely mortal. Though he certainly held tight his grasp on this world when royalty conspired to send him on.

You've probably heard the condensed version of the tale; he was poisoned, shot, beaten, and finally drowned at a dinner party in his honour. Though, lest you get the wrong idea, this was not just how they partied in Russia at the time. No, it was Rasputin's benefactors – who had grown tired of his meddling in political affairs – who decided to put an end to his influence.

Rasputin's friend (or so he thought) Felix Yusopov invited the mystic to a late-night social gathering. To spare you the minutia of detail, much of which is in question, what is known is that Rasputin had been served wine and pastries laced with cyanide. Most accounts agree that he reluctantly imbibed copious amounts of the poisoned wine, though he may or may not have eaten the tainted pastries. Though, for whatever reason, he did not succumb to the toxin, except by way of becoming quite drunk.

Desperation set in and either Yusopov himself, or another conspirator – perhaps Dimitri Romanov – attempted a much more overt act to dispatch the holy man. Rasputin was shot in the left side of his abdomen (some tellings claim he was shot in the back) and he fell to the floor, apparently lifeless.

Thinking they had succeeded, the conspirators made haste to dispose of Rasputin's belongings, and later returned to remove his body, whereupon they found him still very much alive, and attempting escape by crawling up a flight of stairs to the courtyard. Well, this simply would not do, so fellow conspirator Vladimir Purishkevich again tried to shoot him, missing twice and then finding his mark in Rasputin's back as he fled. He took a final round to the head, in what one can only imagine would have been an Oscar-worthy scene, and fell to the snow.

Yusopov, who apparently had been moved to madness by the evening's events, then set upon him with a truncheon, beating him about the head and body until he was finally pulled away. Wouldn't you know it though, but Rasputin was still alive.

Having had enough of this unending assassination, his murderers wrapped him in a carpet and dumped him into the nearby Nina River. Much to the surprise of authorities though, when the body was dredged from the river some days later, there were signs he had yet been alive and had injured his hands trying to break through the ice from the underside.

An unbelievable tale, for sure, but as mentioned, his wounds alone could have been survivable, at least in the short term. Whether he had some immunity to cyanide, or perhaps his hosts administered it incompetently, any single event he suffered that night could not conclusively be deemed fatal on its own. Mostly because none were, until he drowned.

But he's not the only person to have survived grievous injury at the hands of an attacker.

Jumping back into the 21st century, on May 27, 1988, a Suffolk Country, New York police officer named Kenyon Tuthill was ambushed by a crazed man with a shotgun while he sat in his patrol car. Tuthill took a point-blank shotgun blast to the face, which, as you might imagine, removed not only most of his face, but nearly 30% of his head.

Anyone reading this would be right to assume that Officer Tuthill's life ended that night, but what if I told you that he never even lost consciousness?

His attacker fled immediately following the shot, likely assuming the officer dead, but Tuthill was still very much alive. The officer called for help over his CB radio, though with most of his mouth gone, he managed only garbled moans. He underwent nine hours of surgery to save his life that day and his attacker was eventually apprehended and was ultimately sentenced to life in prison.

Officer Keyon Tuthill survived what no one thought was possible; since then he has gone through more than 17 facial reconstruction surgeries, and was a featured story in the documentary Ultimate Survivors: Winning Against Incredible Odds (1991) starring William Shatner.

These two stories from what seem like opposite sides of history, sit in stark contrast to our news headlines. People are killed so often, and sometimes in such seemingly benign ways, and almost always for no good reason. Our bodies are subject to assaults on every level, and at a glance it seems there's no logic involved in who lives and who dies. Rasputin's story typically incites talk of magic, and sorcery, and alchemy, but when a regular guy like Officer Tuthill can survive such massive injury through sheer will, magic seems unnecessary.  One thing is certain though, of all the ways a life can be destroyed, killing each other is the worst.

## The Golem: Friend, Foe, or Farce?

To mark my debut as a Contributing Editor at The Daily Grail, I recently compiled a run-down on the origin of robotics. Surprised as you might be to find that Will Smith's robot archenemies actually owe their existence to 4000 year old Babylonians, the story of developing robotic technology is quite interesting. If I do say so myself.

Of course, the precursors in 2000 BCE weren't called robots; that term didn't come about until 1923. Before that, most examples of complex machinery were called automata, or automatons. Though, that word is a little misleading. Yes, examples of early robots, like Leonardo da Vinci's magnificent mechanical lion, and Muslim engineer al-Jazari's automatic hand washing machines, were called automata, but it had other uses.

Technically, the word – which is derived from the Greek *automatos*, meaning self-acting – refers to anything that operates, acts, thinks, or moves on its own. This is why it applies to early robotic mechanisms. It can be applied to other entities though, as long as they are autonomous. (See what I did there?)

OK, enough shenanigans. Automata come in many forms, and one of those forms is the golem. No, not Gollum. As precious as he is, he's not what I'm talking about. A golem is an artificial man, according to ancient Kabbalistic teachings. But it's more than that. It's the earliest known reference to an android, though, certainly not an android like Lieutenant-Commander Data. Actually, it's kind of complicated.

The first ever golem was actually Adam. Yes, that Adam. You'll recall from your bible studies that Adam was created out of dust by the Almighty. This is the most basic and early definition of the word; to be "kneaded into a shapeless husk". The Talmud says, in Tractate Sanhedrin 38b, that only

God has the power to create life, and all golems were created by God. However, there's a caveat. It also says that a holy man who is devout and whom strives to emulate God would also have the ability, but the life he creates would be but a shadow of that created by God, thus Adam became human, but golems are not (all of the golems referred to below are of the second, lesser form).

The casual reader may read that and think that it means, simply, procreation, or having babies. And while that may be a part of the meaning, there are those who believe it means something altogether different. Here's where things get interesting.

Golems were, or would have been, slaves to their creators, though were also potentially hostile if given the chance. The one sort of defining feature of a golem is the inability to speak. Which is why the word evolved into the modern Yiddish *goylem*, meaning dimwitted or slow.

You can perhaps see why one would wish to create such a monstrous thing. And given that the power to create one was thought to be holy, some of the golem characters in history and literature are…odd.

According to Hebrew teachings, a holy man can create an automaton out of clay. That is, a fully-animate, real, and somewhat spooky golem. The resulting creature – a sandman, if you will – would be, by definition, a lower form of life; incomplete or uncultivated.

To make a golem was deceptively simple. Some of the stories about their creation are, shall we say, difficult to believe (as though this whole endeavour weren't something out of a fairy tale). It's said that specific Hebrew incantations were used, literally, to animate the creature. Something known as an ecstatic experience would be induced through the use of shems (one of the names of God), by applying different Hebrew letters to the clay form. This was done either by writing some variation of the shem on a piece of paper (or parchment) and inserting it into the golem's chest or mouth, or by inscribing the letters onto the surface of the form, often on the forehead.

Some telling's claim that the Hebrew word emet was used (אמת), meaning "truth", and that simply removing the א would change the meaning of emet from "truth" to "death", thereby deactivating the golem. Though, the most famous stories about the creation of a golem differ somewhat.

There are a few relatively famous examples of golems, or more accurately, famous stories of their supposed creation and use. The truth of their existence is doubted, but not by all.

The Golem of Chelm, which was supposedly created by the 16th century Rabbi Ellyahu of Chelm (a city in eastern Poland), came to life either by the hanging of a pendant bearing the word emet around its neck, or with the inscription of the same on its forehead. It's said that Ellyahu created the monster to aid him in heavy work, but (and this depends on the account) the golem began to grow with the passage of time. As it got bigger, he feared it would become so large that it might threaten to destroy the universe. So the rabbi removed the shem, causing the great creature to collapse into dust, but in so doing suffered terrible injury himself during the struggle.

Perhaps the most famous golem story is that of the Golem of Prague. Also in the 16th century, Rabbi Maharal of Prague created his in a desperate effort to protect the Jews of his city from persecution and death at the hands of the Holy Roman Emperor, Rudolf II. Maharal created the creature on the banks of the river Vltava, molding it in the clay of the riverbed. He and his son-in-law performed a ritual, involving the recitation of several Kabbalistic verses while circling the inanimate golem in alternating directions, followed by the inscription of a shem on its forehead.

The great creature awoke and performed its function dutifully, but in keeping with the teachings of the Talmud, Maharal was required to deactivate it every Friday evening, so that it would be inoperable on the Sabbath. For whatever reason, he was unable or simply forgot to do this one Friday eve and, again depending on the account, the creature wreaked havoc across Prague. Maharal brought down his golem in front of the synagogue after a long battle by removing the shem, thereby causing the monster to crumble to pieces.

The remains of the golem were said to have been stored in the attic of the synagogue, now known as the Old New Synagogue in Prague, so that Maharal could reanimate it if ever it was again needed. Many people believe that this story is based in fact, and that the golem's corpse, if you will, was kept in the attic until recent years. Unfortunately, at least for anyone hoping it was, no evidence has ever been found to indicate this was the case.

You might say that these stories are nothing more than fantasy, legend, religious folklore, and you'd probably be right. Let's face it, it's a fantastical thing, and golems have appeared in literature ever since. One of the Brothers Grimm tales involves a golem, and there are many books, plays, and stories that include them by name. Some, however, you might not immediately recognise as golem. Mary Shelly's Frankenstein is a golem tale, though adapted to an age of science and electricity. The Gingerbread Man fairy tale is also a variation on the theme, derived from the Yiddish and Slavic folktales of The Clay Boy. Some even believe that Czech playwright, Karel Čapek, who wrote Rossum's Universal Robots – which is given as the origin of the word robot – actually based his famous play on golem lore (though he denied it).

Modern culture has been influenced by the golem more than most might think. From cartoons and movies explicitly about them, to variations of them appearing in comic books, video games (Minecraft, Assassin's Creed, and even World of Warcraft), and virtually any form of entertainment media since the dawn of their inception. You may know these monsters by many different names, but in reality they are the golem, and depending on your faith in Judaism...perhaps they are even us.

# The Mysterious Celestial Spheres of the Ancient Mughal Empire

Prior to the invention of Google Earth, as some of you may be surprised to find out, when one wanted to see what the planet looked like, or to find a certain faraway place without actually travelling to it, one had few options.  Maps, of course, have always been an option, and you'll recall that they didn't always fit in your phone.

We've made maps for millennia.  It's an art form unto itself.  Cartography is the formal title for it, and as anyone with a love for antique maps can tell you, the variation in form and artistic style is both immense and awe inspiring.

The oldest known map, or cartograph, is difficult to pin down.  There are ancient works of art that could, with some verbal gymnastics, be considered maps, and those date back as far as the 7th millennium BCE, but the first known world map (which is an actual map) dates from 9th century Babylon (BCE).  It is beautiful, though not exactly accurate.

Of course, there are different kinds of maps.  From a technical perspective, there are topological maps and topographical maps – where the former is a very basic drawing highlighting landmarks and routes of passage, and the latter is more finely detailed and concerned with topography, or elevation and land structure identified through contour lines.  There are navigational maps, population maps, faction maps, marine maps, even wind maps.

But yet, there are still more kinds of maps.  Most are concerned with demonstrating relative locations on Earth, but people have been making maps of the stars for almost as long as they've been giving each other badly drawn directions to the corner store.  Celestial maps, as they're called, offer a standardised view of constellations and individual stars, along with their relative position compared to specific points on Earth.

One of the problems with celestial maps, and actually with all maps, is the two-dimensional representation of a three-dimensional object or space. In order to accurately plot locations and show a realistic measure of their position relative to all others, the cartographer must distort the actual shape of either the Earth or the heavens. This, obviously, can cause some problems when one wants to clearly understand the actual relationship between two locations. The answer? Globes!

Globes too, are split into two categories, terrestrial and celestial. The earliest known terrestrial globes date to ancient Greece (6th to 3rd century BCE), though no examples have survived the ravages of time. Celestial globes may have gotten a start much later, possibly as late as 2nd century CE, as a part of the Farnese Atlas, which is a Roman replica of the classical Hellenistic sculpture of Atlas, but depicting him holding up the heavens, rather than the world. Though, since no examples or records of celestial spheres exist prior to, it's not known when or who exactly started the trend. Antique celestial globes are most often made out of metal, usually bronze, and are usually hollow, but are also found in marble and other sculpting mediums.

*The famous celestial globe of Muhammad Salih Tahtawi is inscribed with Arabic and Persian inscriptions, completed in the year 1631.*

In the realm of celestial globes, also known as celestial spheres, there are some spectacular surviving examples, and among those gems are hidden one of history's most vexing puzzles.

In the 1980's, a Smithsonian historian of science, Emilie Savage-Smith, embarked on a journey throughout the middle-east, with the purpose of finding and studying celestial spheres from antiquity. She found a bounty of them, some of the most incredible works of cartographic art and engineering ever made by human hands.

Among those she found there were two distinct types; seamed and seamless spheres. Seamed spheres are, or were, made by moulding two halves of the sphere separately and then soldering them together, ultimately buffing the soldered seam to make a smooth sphere. Then artisans and astronomers would engrave the surface according to whatever specific element of the skies they wanted to depict.

Seamless spheres, however, were another thing entirely; something Emilie Savage-Smith discovered quite unexpectedly.

Up until Savage-Smith made her discovery, it was thought by virtually the entirety of the academic community and by metallurgists the world over, that all examples of hollow metal celestial spheres in existence were of the seamed type. This was owing to the long held belief that creating seamless hollow metal spheres is impossible. It turns out, that isn't true. [1]

One of the earliest examples of a seamless celestial sphere found by Savage-Smith, was found to be from a workshop in Lahore, Pakistan, though she soon found that the technique, described as 'secret wax casting' was widely known by metal craftsmen in Northern India from at least as early as the late 16th century and coming from the Mughal Empire. In fact, some of the workshops identified continued to use the technique up until the 19th century. Though it has apparently now been lost to modern manufacturing techniques.

According to some, the best surviving example of a hollow, seamless celestial sphere is one made by a Mughal metallurgical master and astronomer named Muhammad Salih Tahtawi in 1631. The sphere, known as the celestial globe of Muhammad Salih Tahtawi, is a massive bronze globe adorned with ornate engraving in both Arabic and Persian, as well as numerous pictographic representations of celestial bodies. Its manufacture would have been an immense undertaking, though Salih Tahtawi surely succeeded in creating a masterpiece unparalleled before or since.

The existence of the spheres, which are commonly known as Islamicate Celestial Globes, isn't without controversy though. Aside from the obvious resistance among modern metallurgists to the idea that these objects were created as Savage-Smith asserts, there exists a good deal of misinformation about these spheres, stemming from what appears to be a reluctance to attribute such mastery to the Muslim ruled Mughal Empire. Several people have asserted that the existence of both Arabic and Persian language on many of the surviving examples is explained simply by the suggestion that those features were added long after the spheres were made.  Presumably implying that the spheres themselves were made by a much older culture, perhaps even in a different area of the world.

Bronze casting techniques similar to that which may have been used to create these spheres, such as lost-wax casting, originated approximately 5700 years ago in Israel, but there is no evidence thus far to substantiate such a claim with regard to the seamless spheres.

Circumstantially, it is a well-established fact that Arab and Muslim cultures were responsible for a great many technological and scientific advances throughout the middle-ages and long before.  There seems to be no valid reason to deny that this particular innovation also came from their masters.

Unfortunately, the subject of seamless celestial spheres is little known in mainstream culture, and as such, in the few places it is discussed, the facts are often distorted or even completely made up.  There are those who would like to claim that these magnificent examples of our history are actually OOP-ART (out-of-place-artefacts), suggesting that their origin is related to either a lost pre-historic human culture or aliens.  Though as with most such arguments, there isn't enough information at present to really dive into the discussion.

In any event, once again we are awed by the sophisticated and masterful creations of our forefathers, and once again, our steady march toward modernity has cost us the wisdom of the ages.

[1] Najma Kazi. Seeking Seamless Scientific Wonders: Review of Emilie Savage-Smith's Work. Muslim Heritage. http://www.muslimheritage.com/article/seeking-seamless-scientific-wonders-review-emilie-savage-smiths-work

# The Beast of Gévaudan: A Real Life Werewolf?

Mankind loves its monsters. They litter our writings throughout history, and they continue to be a popular focal point of modern cinema and storytelling. Even in real life, when we have trouble explaining something, we inevitably invent a monster to fill the gap. It's almost a quintessential aspect of human nature. And one of the oldest monsters in our collective lore, is the werewolf.

The werewolf, or lycanthrope (which is Greek in origin) has an ancient beginning. While modern telling's of the legends tend to romanticise and further mythologise the original character, it may have had a more literal beginning.

Common sources suggest that the idea of lycanthropy may actually be pre-historic, but the earliest mentions of such a creature in literature and artwork come from early Medieval conflicts between pagan concepts of animalised warriors and the Christianisation of Europe at that time. Many scholars and historians propose different timelines, evolutionary paths for the legend, and origins for the character, and few agree on any one aspect, other than the fact that depictions of wolf-men type creatures are found as far back as Iron-age Europe (1200 BCE – 1 BCE).

There are several mentions of wolf-men type figures in Ancient Greek literature and mythology, and though the similarities are obvious, it's not generally thought that the Greeks had any part in popular werewolf lore. Most believe that it had a Germanic origin, but there is evidence for the development of such stories from several parts of Europe.

Wherever it began, there's no denying the obsession we have with these stories and characters, and while pretty much every such story you might hear today is without a doubt fiction, there are some cases in the not-so-distant past that aren't so easy to dismiss.

The Beast of Gévaudan is one such story.

Some would argue that Gévaudan's man-eating wolf is not a werewolf story, but if you'll indulge you may be rewarded.

It happened in the former province Gévaudan (now the French department Lozère) in southern France over a period of six years, from the summer of 1764 until the final attacks over Christmas of 1769/1770 – though that's a muddy bit of the story.

The first attacks resulted in the brutal and gruesome death of a young girl named Janne Boulet, and from that point, depending on the source, up to 100 men, women, and children were either killed or seriously injured by the creature.

Sources describe the animal as a massive wolf, or dog-wolf hybrid. It was said to have huge jaws with 42 razor sharp teeth (an important point to refute the claim that it was a hyena), and a red coat with black markings on its back. The animal was apparently as big as a calf, which would make it two or three times as large as any known wolf species.

Embellishments and inconsistencies in the description aside, it's clear that a beast of mythic proportions was loose in the area of Gévaudan, and it had a taste for human blood.

Many of the victims were said to have had their throats ripped out, and several were partially eaten. Of course, the townsfolk weren't about to just sit by while a monster killed them off, one-by-one. Following an attack on a group of men on January 12, 1765, who managed to fight the beast off by grouping together, King Louis XV intervened by sending two of his best hunters to the region. The following weeks were disappointing, and more royal hunters and agents were dispatched to help in the fight. Finally, on September 20, 1765, a wolf of spectacular proportions was killed by François Antoine, the Lieutenant of the hunt. The creature weighed a whopping 60 kilograms (130 lbs), and was 80 cm tall at the shoulder. In his official report to the King, Antoine claimed that they "...never saw a big wolf that could be compared to this one."

Unfortunately for Antoine, the beast he had not killed. Another attack occurred on December 2, 1769, followed by a dozen more deaths.

Of course, every legend needs a hero, and in this case that hero is a local hunter named Jean Castel.  As a member of a hunting party organised by a local nobleman, Castel is credited with killing the creature that had terrorised his village for so long.  Some sources claim that Castel took down the beast with a single shot…a blessed silver bullet.  A silver bullet he personally made for the job of ending their nightmare.

You'll no doubt recognise that bit of lore from modern werewolf stories, and there are those who suggest it's actually an embellishment by previous researchers in an effort to make the stories fit with the legend.

It would have been quite common for people of the time to manufacture their own shot, especially professional hunters, and the notion that silver has an effect on magical creatures was well known in that period.  So it is possible that he did shoot the animal with a silver bullet.

So what was it?

The original stories claim that the beast was slaughtered, and when they opened its stomach they found several human body parts, and since the attacks stopped with the death of this animal, it seems clear that this had been the culprit.  However, there are still some unanswered questions.

For a long time, the Beast of Gévaudan was widely believed to be a real-life werewolf, even by scholars.  Only recently has anyone offered any hypotheses for consideration.  The current contenders are that it was a previously unknown species of wolf that naturally grew to immense size, or that it was a hybrid mastiff-wolf, possibly bred by someone living in the wilderness around Gévaudan.  Others have suggested, as mentioned earlier, that it wasn't a wolf at all, but rather an extant Asian hyena.

The obvious claim that it wasn't any kind of cryptid type animal, but was actually just a pack of regular European wolves notwithstanding, these explanations seem to satisfy logic, even though they greatly disappoint imagination.  The fact remains that to this day, we still don't know what it was that killed up to 100 people in Gévaudan in the 18th century.  We know it happened, and we know that people of the time were convinced that it was a fabled werewolf that terrorised their children and their town.

Perhaps that's all we need to know.

# The Hinterkaifeck Murders and the Devil's Footprints

Six people – a family, well-to-do – murdered one-by-one in their own barn, at the hands of a monster unknown.

Sounds like the plot to a classic horror movie, but that's actually the long-story-short for Germany's most mysterious unsolved massacre; The Hinterkaifeck Murders.

It happened in 1922, in a rural area of southern Bavaria, on the farmstead known as Hinterkaifeck (which means farm beyond or hidden by the woods) in the small town of Wangen (now Weidhofen). On the night of March 31, Andreas Gruber (63), his wife Cäzilia (72), their widowed daughter Viktoria Gabriel (35), along with her two children, Cäzilia (7) and Josef (2), and the new family maid Maria Baumgartner (44), were brutally murdered by persons unknown. All but Josef and Maria were somehow lured to the barn, one-by-one, and bludgeoned to death with a mattock (similar to a pickaxe). The killer or killers then entered the house and slaughtered the toddler and the maid in their beds. It's really not as cut and dried as that may suggest, though.

It turns out that Maria Baumgartner, the maid, was murdered on her first day on the job, in fact she may have been at the farm for only two to three hours before the first murder took place. She had been hired as a replacement for the previous maid, who quit approximately six months earlier, claiming that the farm was haunted.

Some sources cite that a few days prior to the 31st, Gruber had found a trail of foot prints, leading from the edge of the dark forest to the rear of the family home, where they disappeared. They then tell of Gruber and other members of the family hearing strange footsteps in the attic, finding an unfamiliar newspaper in the home, and one of the two sets of house keys going missing in the intervening days. [1] Official sources however, claim that on the morning of March 30th, the day before the murders, Gruber had found that someone had tried to break into his 'motor cottage' (or garage), breaking the lock and disturbing the area outside the feed room. After searching the farmstead for trespassers, he then found a single trail of footprints that led from the woods to the compound. [2]

As mentioned, around 7:30 on the evening of the 31st, all of the adult family members were somehow lured to the main barn and bludgeoned to death. Later autopsies confirmed that a mattock, which was later recovered, had been the murder weapon, and the coroner at the time, noted that the wounds were precise, indicating that whomever had done this was at least familiar with the use of such a tool. After then moving to the main house and using the same weapon on the toddler and maid, they then arranged the bodies in the barn, by stacking them on top of each other, piling hay over them and then covering them with a broken door. They covered young Josef, in his bassinette, with one of his mother's dresses and simply laid the maid on her own bed, covering her with a bed sheet.

Whomever committed this heinous act was apparently quite comfortable with what he/she/they had done, as they stayed in the home for several days afterward, feeding the cattle and having meals in the kitchen, just steps from the corpse of Baumgartner. Neighbours reported seeing smoke rising from the chimney on the following Sunday, and the family dog had been handled and tied up near the barn when the postman arrived on Saturday afternoon. Unfortunately the dog was later brutalised and left for dead with the family in the barn, though it survived.

The gruesome nature of the crime is story enough, but there's much weirdness that goes along with this.

It turns out that paternal responsibility for young Josef had long been in question. Viktoria, who was the official owner of the farmstead, was a rather promiscuous young woman. Several men later came forward, claiming to have known her intimately, but a veritable war went on between Andreas Gruber and their long time neighbour and widower Lorenz Schlittenbauer. It seems Schlittenbauer had also been with

Viktoria, and it was believed that Josef was his son. Schlittenbauer was required to make an alimony payment to the family, and retired any rights he had in parentage. However, during these events Viktoria had elected to marry Schlittenbauer, who was several years her elder, but Gruber objected, and in return allegations of incest were leveled at Gruber, and he was ultimately imprisoned for a year prior to the murders. It's now largely believed that Andreas Gruber was Josef's real father (and grandfather).

The bodies were finally found on Tuesday April 4th, by Lorenz Schlittenbauer and four other neighbours, who had been alerted to something gone wrong by Gruber's absence at church that past Sunday, and the absence of the younger Cäzilia at school on the Monday. They attended Hinterkaifeck late in the afternoon, and following a brief search, found the gruesome scene in the barn.

Subsequent investigation and autopsy saw the corpse's heads removed for study, which were ultimately lost (most likely in the battle at Nuremberg during WWII), and the bodies were buried, headless, in a local cemetery. The farmstead was leveled a few years later, and now a monument stands on the site in honour of the departed.

So who did it? There have been several suspects in the years since, not the least of which was Lorenz Schlittenbauer. His familiarity with the farmstead and the people involved, coupled with the controversy of his would-be son and almost-wife, gave plenty of room for motive. In fact, on the morning of the 30th, Gruber had seen Schlittenbauer at the neighbour's farm while he tracked the trail of footprints, wherein he warned his neighbour of a possible prowler in the area. This could have given Schlittenbauer opportunity to commit an atrocious act, while leaving doubt about who may have done it.

Of course, there's those foot prints. Someone attended the farm, approached on foot from the wood, and apparently never left. Yet no strangers or trespassers were found.

An escaped mental patient was also among the suspects. Joseph Bärtle had slipped away from an asylum at

Günzburg in 1921, and was apparently at large, possibly in the area of south Bavaria at the time.

But what if the foot prints weren't the trail of a man after all? I give you the Devil's Footprints.

Found in February of 1855, following a heavy snowfall in Devon, England, were a strange line of tracks of an apparently two-legged creature with cloven-hoof feet. The strange foot prints were tracked from Exmouth, across the Exe Estuary, to Teignmouth some 40 miles away. There were, at times, large gaps in the trail, where it appeared that the creature had taken flight and then landed further down-field, and it was said that they appeared on rooftops, in gardens and up walls. At the time witnesses attributed the tracks to the devil, hence the name, and though the tracks were studied and diagramed, no one has ever come up with an acceptable explanation for what made them.

Now, the Hinterkaifeck foot prints were never photographed, and the description of the tracks doesn't provide any detail about their appearance. And the Devil's Footprints were never associated with any known crimes or otherwise unexplainable events, but one can't help but see a similarity between the footprints of the Hinterkaifeck Murders and the Devil's Footprints.

Were the family of Andreas Gruber slaughtered by a disgruntled neighbour, a deranged lunatic, or an otherworldly creature who left its calling card in the form of mysterious footprints?

[1] Author not listed. Hinterkaifeck. Armchair Detective: http://armchairdetective.wordpress.com/2009/11/23/hinterkaifeck/

[2] Elfriede Weber Alte Landgerichtsstr. Hinterkaifeck-morde.de (German language): http://www.hinterkaifeck-mord.de/index.html

# The Mystery of the Dyatlov Pass Incident

In the world of weird there are many places that elicit wonder and trepidation.  Aside from the usual haunted locals that many think of, there are many geographical locations that embody mystery, the Bermuda Triangle for one.  Though if there were an election for the capitol of weird, one location would be at the top of the list of consideration: The Ural Mountains.

The Urals, as they are commonly known, are a mountain range stretching north and south through western Russia.  The iconic region ranges from the coast of the Arctic Ocean in northern Russia to the borders of Kasakhstan in the south.  It marks the northern border between the continents of Europe and Asia and is a treasure trove of geological bounty and historical significance.  Its highest peak, Mount Narodnaya, sits at 6,217 feet.

Like many mountainous regions, the Ural Mountains have their share of strange stories and mysteries, but perhaps the strangest story to come out of the area is that of the Dyatlov Pass Incident.

In mid-January 1959, a group of young skiers embarked on a trip into the frozen wilds of Kholat Syakhl, a mountain in the northern Ural range, commonly known as The Dead Mountain (the name Kholat Syakhl means *dead mountain* in Mansi, the dialect of the local Mansi people, and refers to the lack of wildlife on the mountain).  Nine of the ten members of the group hiked into the mountains headed for the slopes of Otorten (one member, Yuri Yudin, fell Ill early on and was forced to turn back).  On February 1st, the experienced group, led by Igor Dyatlov (after whom the mountain pass was eventually named) strayed from their planned route, probably because of poor weather conditions, and found themselves on a high slope of Kholat Syakhl, where they decided to camp overnight.

The events of that night are not well understood, but judging by the aftermath, whatever happened, it was one hell of a night.

Deduced from intensive search and rescue effort and subsequent investigation, the early morning hours of February 2nd 1959 were tragic and eventful. Search efforts, which were initiated on February 20th found an eerie scene. The camp was found on February 26th and it was quickly determined that something bad had happened. A tent was found and it appeared that whomever had been inside had torn their way out of it...from the inside.

Footprints were found leaving the camp, unfortunately, these footprints showed that the hikers were either barefoot, wearing socks or only one shoe. Fearing the worst, search teams upped the intensity of their search and eventually found the first of the group, Yuri Krivonischenko and Yuri Doroshenko, huddled around the base of a large cedar tree just more than a kilometer from the campsite; frozen, shoeless and dressed only in their underwear. The remains of a fire were found nearby, but the -30 degree climate and high winds would not have been staved off by a fire alone.

Next to be found were Dyatlov, Zina Kolmogorova and Rustem Slobodin. All three were found part way between the camp and the cedar tree, in poses that suggested that they were attempting to return to the tent when the elements overtook them.

The other four members, Semyon Zolotariov, Nicolai Thibeaux-Brignolles, Ludmila Dubinina and Alexander Kolevatov, weren't found for another two months. They were eventually found under four meters of snow in a ravine 75 meters beyond the cedar tree. It appeared that these four had survived longer than the previous five, as it seemed that they had taken clothing from their fallen comrades in an effort to survive.

All in all, it seemed apparent that the group had succumbed to the elements, but further investigation generated a host of questions. Why had they left the warmth and safety of their tents? Why had they ventured out without their clothing or shoes?  And why did they not attempt to recover their belongings after leaving the camp?

These compelling questions further drove the investigation and very few answers seemed to emerge.

Early on, the official cause of death for the entire group was listed a hypothermia, but following a more thorough assessment of their injuries, investigators found that two of the group members had fractured skulls and two had severe chest trauma resulting in several broken ribs.  One of the women, Ludmila Dubinina, was missing her entire tongue.  One Dr. Boris Vozrozhdenny estimated that the force required to inflict the chest wounds would have been similar to a severe car accident, but what baffled investigators was the fact that there were no external wounds.  No cuts, scrapes, bruises or other visible trauma.

Theories and speculation began to circulate; some people thought that perhaps an avalanche had struck the camp after lights out, though there was no sign of avalanche damage anywhere near the camp.  Others thought that the local Mansi people had happened upon the group and took exception to them trespassing on their land, and had slaughtered them in the night.  This of course, makes little sense, as the Mansi are not territorial in this manner and have no history of such behaviour, nor could such an attack have accounted for the type of injuries found...no superficial or external wounds.

What is perhaps the most baffling part of the story is the fact that the clothing of some of the deceased showed radioactive contamination at higher levels than would be found naturally.  Though skeptics have correctly pointed out that some camping lanterns use Thorium gas mantles, which can leave radioactive residue on clothing and other items in their general proximity, but who knows if the levels found were consistent with that of Thorium mantles.

Many theories, some more credible than others, have surfaced over the years, but the fact that there were no survivors or eyewitnesses drastically impeded the investigation, which was officially closed in May of 1959.  Investigators cited the "absence of a guilty party" for their laughable conclusion that the group members had died due to a "compelling natural force". The local government also imposed a ban on anyone seeking to camp or even travel through the area, which lasted for more than three years.

Since then, others have come forward with further information, such as another group of hikers who camped approximately 50 kilometers south of the Dyatlov group, claimed to have seen several large orange glowing orbs in the general direction of Kholat Syakhl on that evening. It was also found that the area of the Dyatlov camp was directly in the flight path of intercontinental R-7 missile test launches from the Baikonur Cosmodrome (though it's not clear if there had been any test launches on that night). Some have claimed that there had been an inordinate amount of scrap metal of unknown origin in the vicinity of Kholat Syakhl, and that the Russian government was/is complicit in a cover-up of illegal military dumping.

Few have come right out and said it, but there is the obvious possibility of UFO or alien influence in this case, and though that may seem far-fetched, it remains clear that the members of the Dyatlov group were subjected to a terror that caused them to act in a most irrational and ultimately fatal manner. These experienced hikers put aside common sense and fled the safety of their camp, knowing that the elements were likely to kill them. Why would someone do that, unless the prospect of staying (or returning) was more frightening and dangerous than braving the frigid night?

# The Angikuni Mystery: The Case of the Missing Village

Canadian culture is defined along the terms of our modern society, but there's much more to our identity than our often mocked accent, our maple syrup and our penchant for plaid flannel shirts. Much of our history is rooted in the traditions of our native, or First Nations population. A large part of that population is Inuit, a people whose culture has strong oral traditions and a kinship with the land.

Nunavut, Canada's largest, northernmost and newest territory (distinct from a province only in the way it derives legal authority), is currently home to some 30,000 Inuit. In the 1930's however, and thanks to the Angikuni Mystery, that number is at least 30 people high.

The story, first published in The Danville Bee, a newspaper of the north, and written by reporter Emmett E. Kelleher, broke on November 27, 1930.[1][2] Is seems the day before Kelleher had been regaled by the story of a northern trapper named Joe Labelle, who told of an entire village of Inuit that had gone missing.

As Labelle tells it, he attended the village on the shores of Lake Angikuni – a village he frequented in his travels – expecting a warm welcome, but as he approached the group of elk skin tents he had an odd feeling. The air of the place just gave him "the creeps". Upon entering the small shanty town, Labelle was greeted by two starving and emaciated Husky's, and venturing further, he found a full team of seven dogs that had apparently starved to death.

His calls into the village went unanswered as he began to search for inhabitants. Entering one hut, he noticed cooking utensils and pots, apparently with food still in them. Under a large fur he found a rusty rifle, giving him pause because, according to Kelleher, the Inuit of the time valued their rifles over nearly everything, and leaving such a tool behind would be unheard of.

Examining another tent that had been virtually destroyed by wind, he found the skins of several foxes, ruined by rain and mud, accompanied by another rifle. Rust on the rifles gave him the impression that the village had been deserted some 12 months prior, and judging by the size of the camp, it appeared there had been at least 25 people living there.

## VANISHED ESKIMO TRIBE GIVES NORTH MYSTERY STRANGER THAN FICTION

*The original headlines in the Danville Bee. The included photo was found to be another village entirely, from 1909.*

His mind reeled trying to understand the mystery; where had they gone? Had they simply moved on? Unlikely, with all of the items left behind. Did they all drown in the nearby lake? Also unlikely, as there would undoubtedly be bodies to be found. His next discovery sent chills down his spine.

His thoughts turned to foul play as he stumbled across a grave with a cairn built of stones. One side of the grave had been removed, stone-by-stone and the body was missing. Labelle couldn't imagine a reason for desecrating the grave of a loved one, and he was reminded of an old Inuit superstition.

Eskimo's, of the time and some still today, believe there is an evil spirit that haunts their villages. Tornrark - who has an "ugly man face with two long tusks sticking up from each side of the nose" – is feared by many Inuit, who wear special charms in the hopes of warding him off. [3]

Labelle stayed in the camp for that afternoon, trying to figure out the mystery. *"There were no signs of any struggle. Everything looked peaceful. But the air seemed deadly."*

Following Kelleher's story in the Bee, the authorities were notified and the RCMP were said to have initiated an investigation and search. No one was ever found, nor were any clues as to the reason for their disappearance.

This story caused quite a stir in the area, but soon succumbed to fleeting attentions and was lost to further curiosity. Until it was published in Frank Edward's 1966 book, Stranger than Science. [4] Edwards telling of the story, taken directly from the original article in the Bee, rejuvenated the mystery and sparked some amateur investigation into the details.

Inquiries with the RCMP failed to come up with any evidence of the initial search, and the RCMP officially deny that there was one, and even that there ever was a village of that size in the remote area of Angikuni. Very few records exist regarding Inuit populations in the territories from that period, so it's nearly impossible to empirically prove that the camp existed, let alone that its inhabitants disappeared.

Suspicions of a supernatural influence at work were put forward not only by Labelle, but also by Whitely Strieber in his 1989 novel Majestic, and by Dean Koontz's 1983 horror novel Phantoms. More recently Nigel Blundell and Roger Boar wrote a detailed accounting of the Angikuni Mystery in their 2010 book The World's Greatest UFO Mysteries, where they add to the growing lore associated to the event.

Many modern tellings of the story have embellished the facts, claiming reports of strange lights in the sky, mass grave robbing and over 1000 people having vanished. But the original mystery holds a hauntingly simple narrative, and though Labelle and Kelleher refrained from speculating on the fate of the Eskimos at Angikuni, one's mind does tend to conjure ideas of alien abduction or supernatural mayhem.

We have only Labelle's first hand accounting of the mystery. Having been a trapper for over 40 years, Labelle was of a type of man that isn't known for telling yarns. Many trappers of the time lived solitary lives, seldom coming into contact with other people outside of these small Inuit villages, and beyond an actual member of the village, Labelle was uniquely qualified to understand the nuances of Inuit life and traditions.

The lack of official records on the search and the village does little to sway the belief of those who identify with the mystery. Considering the time frame, it's unlikely we'll ever know the truth of it, but the notion than an entire village of people could disappear, almost overnight, is a disturbing one to be sure.

[1] Kelleher, Emmett E. (1930-11-30). "Vanished Eskimo Tribe Gives North Mystery Stranger Than Fiction". The Bee.

[2] Newspaperarchive.com, The Danville Bee – November 27, 1930: http://newspaperarchive.com/danville-bee/1930-11-27/page-7

[3] Colombo, John Robert. Ghost Stories of Canada. Dundum (2000)

[4] Edwards, Frank. Stranger than Science (5th printing ed.). Bantam Books Paperback (1968). pp. 18–19

# The Clapham Wood Mystery

The world has a long history of forests with apparently supernatural qualities, from the legendary Black Forest of southwestern Germany, where every manner of creature, from werewolves to sorcerers, are said to originate.  To England's haunted Wychwood Forest, possibly the most haunted forest in Britain.  To Japan's Aokigahara Forest, also known as the Sea of Trees, where more than 500 suicides have been reported since the 1950's.

There's something vaguely romantic about a mysterious wood, like Sherwood Forest of Robin Hood fame, which was known as a haven for everything evil.  The idea that a forest holds wicked truths and supernatural power seems to speak to us in a primal way.  Ancient stories perpetuate the myths of such haunted places, and the anonymity of a ghost story involving such a vast area seems to give license to more than a few urban legends.

One such instance of strangeness is Clapham Wood, where the aptly named Clapham Wood Mystery has been confounding paranormal researchers for decades.  Located in West Sussex, England,

Clapham Wood stands to the north of the small village of Clapham.  Historically, Clapham has been an archetypal English village, one that's been around, likely, since Saxon times.  Over the last 300 years it has remained largely hidden from the outside world, except for the last four decades that is.

Perhaps the most mundane feature of Clapham Wood is it's abundance of UFO sightings.  Since the 1960's there have been hundreds of strange sightings both in the woods and in the village itself.  Of course, to say that UFO sightings are mundane may give you the wrong impression.  The area has been the focus of a great deal of UFOlogical study, and has hosted a number of sky-watching vigils over the years.  In addition, people have reported unexplained nausea and the distinct feeling that they were being followed.  Reports of a strange grey mist appearing suddenly on pathways throughout the wood, and instances of ghostly forces pushing hikers and dog-walkers have been known for years.  And studies using a Geiger counter showed a slightly higher level of radiation in the area.

The reason for the moniker 'Clapham Wood Mystery' though, is its ties to Satanic Cults.  In 1975, residents of Clapham were plagued by the illnesses and disappearances of beloved pets.

Three cases of which were covered widely by the press.  They told of two dogs that went missing without a trace and a third that suffered a mysterious paralysis.  The son of Peter Love, while walking their family chow in the wood, watched as his dog ran amongst the trees of the forest and disappeared, never to be seen again.  The following week, farmer John Cornford's collie disappeared in the same place.  A third dog, a golden retriever owned by Mr. E.F. Rawlins was found partially paralysed after running into the woods, the cause of which was never determined and which eventually led to its being euthanized.  This was only the tip of the iceberg however.

In April 1972, the body of an unidentified young woman was found in Clapham Woods.  The case was investigated by Police Constable Peter Goldsmith, among others at his detachment.  In June of the same year, Goldsmith himself disappeared.  Goldsmith, a former Royal Marine Commando had last been seen walking across the rolling Downs (grassy chalk hills) near the town's 13th century church.  Despite intense investigation and a wide search of the entire area, his body wasn't found until six months later, on 13 December, amid a thick patch of bramble less than a half mile from the location where the girl's body had been found.  No suspect was ever identified, nor was a cause of death.  And it doesn't end there...

In July 1975, pensioner Leon Foster disappeared and was subsequently found three weeks later, by a couple who were searching for a horse in the wood, a horse that had also gone missing under mysterious circumstances.  Next, on

Halloween of 1978, the vicar of Clapham, the retired Reverend Harry Snelling went missing.  His body was found three years later, by a Canadian tourist.  Again, no cause of death could be identified.

And in 1981 a homeless schizophrenic named Jillian Mathews disappeared.  When her body was found in the forest police determined that she had been raped and strangled. [1]

While no culprit was ever brought to account for these heinous crimes, this doesn't mean that there aren't a few theories about who was responsible.  Paranormal investigator Charles Walker, who was looking into the strange goings-on in and around Clapham, received a telephone call from an unknown man.  Described as "well-spoken", the man claimed to have inside knowledge of the cause of the strange happenings of Clapham Wood.  A very spooky late-night rendezvous was arranged inside Clapham Woods, at the "Cross-Roads".  When Walker arrived at the agreed location, he could find no one there, until a voice called out from the darkness of a nearby bramble claiming that it would be dangerous for both of them if Walker had seen the identity of the mysterious tipster.

The concealed informant explained that he was an initiate of the Satanist cult called The Friends of Hecate.  He claimed that the woods were used for their monthly meetings, and that the missing animals had been used as sacrifices to Hecate.  He made no mentioned of the deaths of the Constable, or the Reverend or any of the others, but Walker drew a connection between all of the events.  The mysterious man then issued a dire warning, saying:  *"There are people in high places involved, holding positions of power and authority, who will tolerate no interference.  We will stop at nothing to ensure the safety of our cult."*[2]

This warning came close on the heels of reverend Snelling's disappearance, barely a week before.

The Friends of Hecate are reportedly an occult sect that worships Hecate, an ancient pagan goddess said to be associated with cross-roads, fire, light, the Moon, magic, witchcraft, knowledge of herbs and poisonous plants, necromancy, and sorcery.[3]

Typically worship of Hecate finds it roots in a re-evaluation of ancient Greek mythology, and is not readily attributed to Satanism. Nonetheless, The Friends of Hecate laid claim to the mysterious woods of Clapham for a period of at least 20 years, allegedly holding monthly ritualistic gatherings, at which a blood sacrifice was always required. It is believed that the group used the Clapham Woods until they were forced to find an alternate location after the Great Storm of 1987 damaged large tracts of the forest.

Apparently as a result of all the suffering and death that have occurred in and around Clapham Woods, many paranormal investigators today believe the forest to be among the most haunted places in England. When one considers the morbid history of the area accompanied by the reported UFO activity, one is hard pressed to retain scepticism of this assertion.

Something strange calls the woods of Clapham home.

[1] Mathew's death is considered by some to be independent of the Clapham Wood Mystery.

[2] Cawthorn, Nigel. Satanic Murder. AuthorsOnline.co.uk http://www.authorsonline.co.uk/book/166/Satanic_Murder/sa mple/

[3] d'Este, Sorita & Rankine, David. Hekate Liminal Rites, Avalonia, 2009.

# Fortean and Weird Stuff

## Did a Woman Stay Awake for 30 Years? Insomnia Explored

Have you ever had insomnia?  Sure, most people have at some point in their lives.  It's one of those things that gets generalised and used to mean "I didn't sleep so well last night."  And while that's technically an example of insomnia, those of us who truly suffer from extended bouts of sleeplessness have little patience for that diminishing attitude.

I suffer from chronic insomnia.  I have to take a sedative to get any sleep at all, and that usually works fairly well, though there are times when I curse my defective brain.  Actually, I do that often, but usually for different reasons.

My current record for being awake – and I mean totally, completely, unmistakably awake – is just over 90 hours.  It wasn't a fun experience, I assure you.  Luckily, my daily schedule isn't so demanding that I couldn't cope with the blurred vision, constant headache, and lack of focus, not to mention a deeply unpleasant disposition.  I really can't imagine what might have happened had I not finally found relief in medication.  Unfortunately, for me and you, imagination isn't really necessary to find out.

Sometimes (incorrectly) called total insomnia, there is an extremely rare disease that causes those who suffer through it, to stop sleeping...forever.  That's a little misleading though, since for those poor people, forever is only about 18 months.  That's because Fatal Familial Insomnia (FFI) is terminal and there's no known cure or remedy.

Before we get to an explanation of what this is, just take a moment to think about that.  No sleep, ever again.  Conscious, aware, awake; twenty-four hours a day, seven days a week, 365 days a year.  Monotonous is a word that comes to mind, though it's wholly inadequate to describe such an experience.  Tortuous might work better.

I'm reminded of one of my favourite Creepypasta short stories; The Russian Sleep Experiment.  If you've never read it, you really should.  It will give you chills.  The original author of the story is unknown, but that's OK, this tale has a life of its own.  For fear of offering unsolicited spoilers, I'll just say that the subjects of the Russian Sleep Experiment fared no better than those who suffer from FFI, and arguably, fared far worse.

FFI is a neurological condition caused by a misfolded protein in the DNA of the afflicted, of which there have been only about 100 cases.  That protein, called a prion protein, is known as PrPSc (PrPC in non-FFI subjects).  Essentially, the prion form of the protein causes a change in certain amino acids – due to the protein strand folding incorrectly – which, when combined with other genetic markers, then affects the brain's sleep centers.  FFI is genetic, and therefore hereditary, but there is an even rarer form known as Sporadic Fatal Insomnia (sFI) that occurs spontaneously, the cause of which is not understood.  You may wish to know that PrPSc is the same protein that's responsible for bovine spongiform encephalopathy, also known as Mad Cow Disease.

You may be thinking that this is all well and good; interesting yet disturbing, but not exactly in keeping with my usual topics.  Well, here goes…

There have been, over the years, many stories – urban legends if you will – describing the incredible experiences of people who remain awake for years, decades sometimes.

The famous UK magazine Fortean Times covered the story of Ines Fernandez in 1975; a Spanish woman of 57 years (at the time), who claimed to have been awake for 30 years. [1] They told the story of the day following the last time she ever slept, suggesting that she suffered some kind of physical neuronal trauma (apparently caused or triggered by a yawn) and then simply never fell asleep again.

There's also the Vietnamese man, Ngoc Thai, 71, who claims he hasn't slept for more than 41 years. His insomnia began after he suffered a mysterious fever in 1973, and according to all sources, he suffered no ill-effect, outside of the fact that he never sleeps.

Thai and Fernandez are special cases. As mentioned, Thai suffered no diminished capacity from his insomnia, in fact he continued to work on his own farm throughout his life. Fernandez is said to have suffered no other symptoms either, beyond a self-described depression following the death of her husband, which she blamed on the loneliness of the long nights.

This shouldn't be possible. Sleep deprivation causes – almost universally – fatigue, clumsiness, weight loss/gain, diabetes, decreased cognitive function, headaches, hallucinations, depression, hand tremors, seizures, mania, and ultimately death, all over a period of months. These symptoms and effects are well documented and studied, so how is it that Ines Fernandez and Ngoc Thai – and presumably others – could remain awake for decades, yet suffer no ill-effect.

There are some explanations. In both cases, the associated literature and verbiage across the internet, is that they were thoroughly examined by doctors, all of whom came up empty in their diagnosis. You'll note that both are from areas of the world not exactly known for the competence of their medical institutions. Even so, one would think any doctor could tell if a person is awake or not.

The popular answer in this is that all of these cases are the result of some supra-natural ability or skill that's being accessed by the patient, and some people in new age circles have held them out to be spiritually significant – Ngoc Thai in particular is something of a celebrity in Vietnam.

There is a more likely explanation though, as I'm sure you guessed. It's a smidge more mundane than having the superpower of total insomnia, but it fits relatively well. It's called sleep state misperception (SSM), which as may be apparent, describes people who mistakenly perceive periods of sleep as wakefulness.  Basically, they really do sleep, but they just don't realise they did.

That almost seems like an April fool's joke from the medical community, but it really is the cause of many of these claims.  It's classified as a sleep disorder through the International Classification of Sleep Disorders (ICSD), though it's more of a psychological condition than a physical ailment.  The key to SSM is that those who suffer with it will earnestly claim that they haven't slept, or have slept very little, but during sleep studies, they show normal sleep patterns.

The upshot is that SSM doesn't come with the terrible side-effects of FFI, such as death, which is handy.  Those with SSM often report depression, though the causal relationship between the two disorders isn't as solid as most think.

So, to spell it out, the woman who hadn't slept in 30 years, probably slept a lot more than she thinks.

[1] The Woman Who Hasn't Slept in 30 Years. Trivia-Library.com: http://www.trivia-library.com/b/mystery-and-strange-phenomenon-from-the-fortean-times-part-3.htm

# Toad-in-a-Hole: The Weird Story of Entombed Animals

There's some pretty weird stuff in this world, and much of it was chronicled by my personal hero Charles H. Fort. His work spawned an entire field of study, though there are those who balk at the suggestion that Forteana is anything but pure pseudoscience. I disagree, clearly.

Forteana covers a wide range of phenomena; from teleportation, to weird rain, to strange lights in the sky. In fact, all of the phenomena covered under the terms occult, supernatural, and paranormal can be considered Fortean. There's a lot to read.

One such phenomena – one I don't think Fort himself ever investigated, but I could be wrong – is Entombed Animals.

No, that's not some elaborate funeral rite for dearly departed pets, it's the strange occurrence of animals – usually frogs, but sometimes small lizards and other animals – being found encased inside rock (and other material). Strange indeed, but it gets stranger.

Of the 200-some reported cases of entombed animals, the vast majority found the animal alive…fully encased inside a void within the rock, or whatever material was involved. This, obviously, raises a few pertinent questions. Chiefly, how'd they get in there, and how are they still alive?

As mentioned, there are more than 200 known cases of entombed animals throughout Europe, North America, Africa, Australia and New Zealand since the fifteenth century. Some more notable than others, for example:

In 1761, physician to Henry III of France, Ambroise Pare, reportedly found in a large and hard stone, a "…huge toad, full of life and without any visible aperture by which it could get there."

Scientific American published an article in 1864 describing the experience of a silver miner named Moses Gaines, who found a toad living inside a large boulder. [1] It was said that the toad had strange eyes that were much larger than those of other toads of the same species.

68 live toads were found inside a tree trunk in South Africa in 1876: "They were of a light brown, almost yellow color, and perfectly healthy, hopping about and away as if nothing had happened. All about them was solid yellow wood, with nothing to indicate how they could have got there, how long they had been there, or how they could have lived without food, drink, or air."

Surprisingly, some relatively famous names have been connected with the toad-in-a-hole phenomenon, as it's sometimes called (not to be confused with the English sausage casserole dish). Benjamin Franklin most notably, who wrote An Account of Toads Found Enclosed in Solid Stone [2], as well as Charles Dickens and Julian Huxley.

Predictably, modern science claims that the entire idea is nonsense, and it's hard to disagree, but for the number and quality of the reports. The journal Nature concluded, in 1910, that the explanation was quite simply "a frog or toad is hopping about while a stone is being broken, and the non-scientific observer immediately rushes to the conclusion that he has seen the creature dropping out of the stone itself." That is, admittedly, a little flimsy.

Renowned paleontologist William Buckland – the man who first described and named the Megalosaurus – conducted experiments in the 1820's, wherein he placed toads into carved chambers in limestone and sandstone blocks and buried them in his garden. A year later he dug them up and, not surprisingly, most of the toads were dead and decaying. But not all of them; an undisclosed number of the toads encased in limestone survived for a full 12 months, cemented inside stone and buried underground. Buckland then reburied the surviving toads and found, another 12 months later, that all had perished. [3]

That's a fairly incredible result, and there are many theoretical explanations for why those toads survived under those conditions. Such as micro-pores in the limestone allowing small amounts of oxygen and water into the void, and various forms of suspended animation (which some frog species are known to employ during hibernation).

This of course says nothing about how the toads and other animals in the 200 plus accounts of toad-in-a-hole got into the rocks, trees or wherever they were found. This defies logic like a teenager defying his parents. Are we talking some kind of teleportation? Macro-osmosis? Magic?

It's conceivable that fertilised toad eggs could have gotten trapped in sediment and eventually ended up embedded inside certain kinds of stone, such as limestone. But limestone forms over millennia or at the very least many centuries; wouldn't that mean that the toad, or its egg, was live or viable inside the stone for hundreds or thousands of years? And what about other kinds of animals?

You can see why the scientific establishment just shakes its proverbial head and walks away amid a cloud of frustrated exhaustion in the face of our tomfoolery. These reports exist though, and are of the sort that aren't easy to dismiss, so we're in a bit of a quandary.

There hasn't been a recorded occurrence of an entombed animal since about 1980, so DNA analysis and other modern scientific methods have never been applied to this phenomenon, outside of an examination of the anecdotal accounts. Thus, many questions about specific species, age, and special genetic conditions are left unanswered.

And speaking of anecdotal accounts, I leave you with the following incredible but unconfirmed (and highly suspect) report coming to us from the Illustrated London News of 1856:

"Workmen were digging a railway tunnel through a layer of Jurassic limestone when they were startled to find a large creature stumbling out of a recently split boulder, flapping what looked like wings, and croaking. It died immediately. The creature was identified as a pterodactyl by a local paleontology student who recognized the characteristic features of the extinct reptile. The stone in which it was found was consistent with the time period in which pterodactyls lived and formed an exact mold of the creature's body."[4][5]

[1] Age of the Human Race: Scientific American 11, 228-228 (8 October 1864) doi:10.1038/scientificamerican10081864-228a

[2] Benjamin Franklin. "An Account of Toads Found Enclosed in Solid Stone." In The Works of Benjamin Franklin, ed. Jared Sparks. Volume VI. 1882. 441–442.

[3] Curiosities of Natural History. Blackwood's Edinburgh Magazine – March 1858 pg.360

[4] Johnathan Whitcomb. Tunnel Pterodactyl of 1856. Live Pterosaurs: http://livepterosaurs.blogspot.ca/2010/04/tunnel-pterodactyl-of-1856.html

[5] Living Fossils – Weird Encyclopedia: http://www.weird-encyclopedia.com/living-fossils.php

# Bigfoot By Any Other Name…Is Just A Man? The Origin of the Word Sasquatch

I have an affinity for, or an affection for cryptids, specifically for Bigfoot.  If you follow me on twitter, you probably know me best by my avatar, which is Harry the Bigfoot from the 1987 hit movie Harry and the Hendersons (one of my all-time favourites).  I should be clear though, I'm not a believer.  That is to say that I'm decidedly undecided on the reality of Bigfoot, but I truly love the mythology and traditions surrounding the concept.  I would be overjoyed should some field researcher, or 'squatcher', bring the world undeniable evidence of the existence of this giant, hairy wild-man of the backwoods in, not only the Pacific North West of the United States, but also in other countries the world over…but I'm not holding my breath.

Bigfoot is perhaps the most famous mythological creature in human history, and there are many people making it their life's business to seek out all information and knowledge on the subject, and to find evidence of this elusive beast, or beasts as the case may be.

But there's an aspect of the Bigfoot phenomenon that a great many people don't know, and it's an issue that is formative to the entire mythology.  We all know that the name of Bigfoot, Sasquatch – which is used by most researchers because is seems to lend a small degree of credibility to the search – is actually a Native American / First Nations word meaning hairy wild-man, but do you really know the story behind that name?

The word Sasquatch isn't technically a Native word, it was coined by Canadian teacher and Indian agent J.W. Burns in the 1920's.  Burns taught for many years at the Chehalis Indian Reserve (No.5&6), which sits on the banks of the Harrison River near Vancouver, British Columbia (between Deroche and Agassiz).  That reserve houses the Chehalis First Nation band of Sts'Ailes people, who were almost wiped out by early European settlement of the area, and who have rebounded from the time of the horrible Residential Schools to a population of over 1000 band members.

Burns was, arguably, obsessed with the Indian tales of giant hairy wild-men, and he wrote extensively on the encounters that were shared with him by tribal elders and travellers. It was through his writings that the word Sasquatch was brought into mainstream culture. He wrote an article for the popular Canadian MacLean's Magazine (April 1929 issue), in which he used the term frequently and since then it's been a household name.

The problem is, the word Sasquatch was most likely a mistranslation. That word doesn't actually exist in the oral traditions of the people in question, nor in any other Native culture in North America. The hairy wild-men of which Burns was a fanatic, apparently do exist, whether as a reality or as a fairy-tale, but they were known by many different names, depending on the specific tribe or band being referenced. It's generally thought that Burns confused the spelling and pronunciation of the Chehalis word 'sasqac'. This word means beast, but there are other contenders for the correct etymological originator, such as 'sokqueatl' and 'soss-q'tal', both of which mean wild-man, according to cryptozoologists Loren Coleman and Jerome Clark. [1]

It isn't necessarily that Burns made a mistake, or misunderstood what was being said, some think he deliberately combined several words in an effort to make an umbrella term to cover all of the various languages he was working with, but it's generally accepted that he did make the word up, for whatever reason. And as such, we now have a blanket term, a household name for the creature or creatures that have been known to Native American and First Nations people for centuries.

There's more to this, though, and it gets a bit weird.

World famous researcher and author Gian J. Quasar, renowned for being the authority on the Bermuda Triangle, and the creator/editor of The Bigfoot Blatt, has a slightly different theory. Quasar says that Sasquatch has a completely different meaning, one you won't be expecting.

In the first issue of The Bigfoot Blatt, of which there appear to only be two issues, Quasar expanded on a theory subtitled Lingua Fanca [sic] – Chinook Trading Jargon: A Skoocum Language, wherein he outlined the etymological origins and evolution of several words, apparently of the Chinook language. He explains the origin of the word skoocum, suggesting that it began as the name of a greatly feared henchman of the Klikatats Indian band, who was known as the Casanov Skoocoom (or the henchman of Casanov, who was the chief of the tribe). Skoocum is now used to describe someone who is good or excellent, or 'cool', and Quasar says that's because the Casanov Skoocum was such a good murderer.

Quasar notes that the words in question are considered lingua franca (as he apparently tried to signify in the subtitle, listed above), or working languages, and are used to make communication possible between peoples who do not share a common mother tongue. And it's through this process that he claims that Sasquatch actually means Saskahaua George.

Quasar claims that Sasquatch came about as an alternative word meant to describe long haired wild-men of King George, or white men if you prefer. He says that Indian warriors were known as sawash (or siwash), but they didn't want to refer to non-Indian's by the same term, so saskahaua was invented. "Saskahaua George comes down to us as "Sasquatch" because the Indians seldom liked to refer to them as sawash (siwash a century ago). That implied they were Indians. But this is something that offended the Indians."[2]

By implication, Quasar is saying that Burns coopted saskahaua, which ultimately became Sasquatch, which has now gone down in history as the Native word for giant, hairy wild-men, or Bigfoot.

Now, despite Quasar's standing as a relatively respected researcher on the Bermuda Triangle phenomenon, he doesn't appear to be a linguist, and his connection, if any, to Native American / First Nation customs is entirely unconfirmed. That and the fact that the Chinook peoples are not related to the Chehalis people (though they were neighbours, geographically), makes his theory a little sketchy. It's an interesting thought though.

What if the word we're all using to identify a huge, hairy, possibly mythological cryptid actually means white-man-of-King-George? I doubt Quasar is going to convince anyone to give up the word now, but it does pay to understand just where our linguistic icons really come from.

[1] J. Clark & L. Coleman. The Unidentified & Creatures of the Outer Edge. Anomalist Books, 2006. ISBN 1933665114

[2] Gian J. Quasar. Lingua Fanca – Chinook Trading Jargon: A Skoocum Language. The Bigfoot Blatt – Issue 1, page 2. http://www.bermuda-triangle.org/html/the_bigfoot_blatt_issue_1_page.html

## Death Explored: The Hanging Coffins of Sagada

Mankind has had a long-standing obsession with death.  We have philosophised over its implications and the possibilities brought about through its transition.  Whole canons of literature have been written, expounding on its beauty, its finality and its mystery.  All of our religions are founded on the idea that something exists after death and they each prescribe the sometimes very elaborate and ritualistic methods we use the world over for the treatment of the deceased.

Archaeologists and historians assert that the custom of burying our dead is the oldest religious rite in our history.  We've been committing the remains of our loved ones to the Earth for at least the last 100,000 years, and probably quite a bit longer.  In fact, the habit of burying the dead is considered a primary indicator for measuring the development of primitive populations.  Though we aren't alone in this habit, both chimpanzees and elephants are known to bury or cover their fallen companions, though none do so with such pomp and circumstance as we.

The earliest evidence of ritual interment was found in a cave called Es Skhul in Israel, on the slopes of Mount Carmel.  Discovered somewhere between 1929 and 1932, the find at Skhul proved fruitful in anthropological terms, bearing 10 full human skeletons – seven adults and three children.  The bones were covered in a red ochre and were accompanied by a variety of grave goods, from specific animal bones, like a boar's mandible, to marine seashells.  The remains and artefacts were dated to between 81,000 to 120,000 years old, and further study revealed that the skeletons were of a people who were the first to have the anatomical tools necessary for speech.

The Skhul remains, as well as the Qafzeh remains found in a cave in lower Galilee, which were found around the same time and were dated to approximately 92,000 years, are strong evidence that ritual burial became a feature of the early spiritual life of our ancestors over a period of decades and even centuries before 100,000 years ago.

Since then, we've complicated and serialised the process of dying. We've built monuments to commemorate it and churches to celebrate it. Today in America, death is a $15 billion industry, not including gravestones and monuments. Of the over two-and-a-half million people who die in the United States annually, a number that rises every year, more than half of them end up in a coffin or casket that ultimately gets buried in a cemetery plot. The rest are either cremated (and sometimes also buried) or are used for scientific or educational purposes.

The same isn't necessarily true for the rest of the world though. While every human culture on Earth holds to some type of after-death ritual, such as burial or cremation, not all dead bodies end up six feet under.

Notwithstanding the various other methods of disposing of the dead, such as burial at sea and even pure neglect, some cultures have taken to stringing their dead up and hanging them from the side of mountains.

Enter the Mysterious Hanging Coffins of Sagada.

Sagada is a region in the Mountain Province of the Philippines, and in this area, as well as parts of Indonesia and China, local peoples have for centuries been hanging their deceased loved ones in ornate coffins from the mountain cliff sides. This is a tradition that originates with the ancient Bo people of southern China, and is still practised today.

The coffins are traditionally carved out of a single log or piece of wood, often by the deceased during their lifetime. They are decorated ornately and painted, often in bright colours, and ultimately hung off of cliff faces, or in cave openings. Sometimes they simply sit on rock outcrops or are suspended by beams. Families have plots of rock face with a line of ancestors hung one above the other, though not everyone qualifies for this special type of burial. Depending on the region, these special burials were reserved for tribal elders or persons of spiritual significance, and others were required to have both children and grandchildren, as burial in this manner was/is thought to be of spiritual benefit to the younger generations.

It's thought that this tradition may have begun as a way to reduce the risk of predation by animals and is a product of the terrain. Some of the coffins that can be seen on the cliffs at Echo Valley in Sagada are centuries old.

The funeral rites involve elaborate processions, often with the family carrying the corpse to the coffin at the hanging site. In these cultures, bodily fluids from the deceased were considered to be sacred and to contain the talent and luck of the deceased. If the procession carrying the corpse came into contact with such fluids, it's thought to be a good omen.

The deceased would often be dressed in family colours and would be interred with spiritual belongings, and would traditionally be forced into a fetal position before the coffin was sealed. The dress and position, combined with the hanging of the coffin was believed to bring the deceased closer to heaven and offer a good vantage point from which to watch over their survivors.

The hanging coffins of Southeast Asia are a sight to behold, as is evidenced by the pictures, but it's said that those who view these vertical graveyards in person are forever changed by the experience. From the perspective of Western society, these traditions may seem odd or even backward, but like any funerary ritual, they are deeply spiritual and engender powerful emotions for those involved.  Our tradition of burying death under our feet could be said to create an out-of-sight-out-of-mind type of attitude toward our departed loved ones, but  no matter what your personal leanings, there's clearly no right or wrong way to say goodbye to those who have moved on to whatever awaits us all.

# When it Rains, it Pours: The Story of Weird Rain

Have you ever wondered where the term "it's raining cats and dogs" comes from? We live in a world where all manner of things is known to fall from the sky. Most commonly, of course, we're talking about water, in its various forms. Mist, fog, drizzle, showers, rain, torrents, hail, sleet, snow, fish, frogs, worms, spiders and unidentified chunks of meat.

It may not be as exotic as the rain of diamonds and glass as has been suggested might fall on certain exo-planets identified by NASA, but our little blue ball sends some fairly weird stuff hurling from the heavens.

In answer to the opening question, no one really has any good idea where it came from, but there are some theories. The leading explanation stems from the thatched roofs of merry old England, where small animals, such as cats and waterspoutdogs, would borough into to the insulating material of the roof for shelter and would fall out during heavy rains. This has never been confirmed, and there are other theories that compete for the 'most plausible' position.

As mentioned above though, there is a history of some pretty strange stuff raining down on our streets and heads, besides the usual Dihydrogen Monoxide. There are more than 17 documented cases of small animals (and other things) falling from the sky since 1861, and no doubt many more prior to that time. In fact there are depictions of fish rain from as far back as 1555 and earlier.

One of the strangest cases of non-rain rain, is that of the Cosmic Meat from Olympia Springs, Bath County, Kentucky (USA) on March 3, 1876.

As reported in the New York Times on March 10, 1876, a woman named Mrs. Crouch was in her yard, making soap when what appeared to be small chunks of meat started falling from the sky.[1] She described them as resembling large snowflakes, but some of the pieces were said to be as large as four inches cubed. Eyewitnesses claimed that the meat looked like beef, though two men who either bravely or foolishly tasted it, said it was either venison or mutton.

The weird thing, as though meat rain isn't weird enough, is that according to Mrs. Crouch, the sky was perfectly clear. Several theories were passed about, and through analysis of the meat by a number of doctors, it was said that the most likely culprit was vultures or buzzards. The doctors found that the meat was a combination of lung tissue, muscle tissue and connective tissues with cartilage, most likely being of equine origin. [2] Officials believed that buzzards had feasted on a freshly dead horse nearby, and while flying overhead, one of the birds disgorged itself (threw up), and as they are apparently known to do, the rest of the flock followed suit, ultimately casting their dinner down on the head of Mrs. Crouch.

The meat rain covered an area of approximately 5000 square yards, which raises the question; just how many buzzards would be required to achieve such coverage? And would a large number of birds be able to fly high enough so as to be invisible to the naked eye from the ground?

The buzzard theory was the most plausible explanation of the time, though there was really only one competing idea, so calling it the most plausible doesn't say much. That other theory was forwarded by American journalist, humourist and author William L. Alden, wherein he claimed that cosmic meat floated about in outer-space with some abundance, and would, occasionally, fall to Earth in the manner of meteorites. [3]

Fresh meat isn't the only weird thing to fall from the skies though, according to Wikipedia, as recently as September 12, 2013, fish were reported to have rained down in Chennai, which is the capitol city of Tamil Nadu, India. Frogs too, are known to fall from the heavens. Theories as to how this happens range from the suction of waterspouts which then fuel storms over land, bring small fish and other animals from lakes and other bodies of water, eventually depositing them far from their homes as the storm loses its momentum. Others have suggested that fish eggs are taken up by these waterspouts, wherein the eggs hatch in the clouds, resulting in baby fish raining down. Though it seems unlikely the eggs could stay airborne for a sufficient period of time for this to be true.

Recent headlines told the story of another strange weather phenomenon occurring in India. This time it wasn't an animal per se, but red rain. As a part of an ongoing phenomenon, the last event occurring as recently as December 2012, residents of Sri Lanka and other parts of the Indian subcontinent found themselves in the midst of a strange series of rain storms that would turn their clothes pink. In contrast to the above, this was rain, in that it consisted of water, but…it was red. Other colours have been reported over the years, from yellow to brown to green, and there has been a good deal of debate as to what exactly it is.

Early theories suggested that perhaps it was some kind of bacteria in the water, perhaps picked up from local waste waters or the Indian Ocean, which is currently the official explanation. Others said perhaps, it was fine meteor dust trapped in the upper atmosphere being condensed by the action of the rain. Close analysis didn't bear that theory out, however.

In January of 2006, two physicists from the Mahatma Gandhi University in Kottayam India, published a paper in the journal Astrophysics and Space Science, which suggested that the red rain was in fact caused by extraterrestrial biological cells brought to Earth via comets and meteors. [4] This theory, of course, is hotly contested, but it remains a part of the discussion.

So, as mentioned earlier, no one really knows where the phrase "it's raining cats and dogs" came from, but if experience is anything to go by, perhaps it harkens to an actual event where it really did rain cats and dogs. Maybe not though.

[1] Author unknown. Flesh Descending in a Shower. New York Times, March 10, 1876.
http://query.nytimes.com/mem/archive-free/pdf?res=FB0914F9355B127B93C2A81788D85F428784F9

[2] Mysterious Showers of Meat. Scientific American –
Supplement 2,437. July 22, 1877
http://rr0.org/time/1/8/7/6/07/22/MysteriousShowerOfMeat_
ScientificAmerican/index.html

[3] Alden, William L. Domestic Explosives and Other Sixth
Column Fancies (From the New York Times). Lovell, Adam,
Wesson & Company, 1877.  Page: 50-52
https://openlibrary.org/books/OL20463677M/Domestic_Explosi
ves_and_Other_Sixth_Column_Fancies_(From_the_New_York_
Times)

[4] Godfrey Louis, A. Santhosh Kumar. The red rain
phenomenon of Kerala and its possible extraterrestrial origin.
Astrophysics, Space Sci. 302 (2006) 175-187. arXiv:astro-
ph/0601022

# EMF Meters...Why?

Of the many posts on this website, some are more popular than others. Some topics seem to strike a chord with readers, or perhaps it's my treatment of those topics that stir the hearts and minds of the few who happen across my work.

One of those popular posts has been what could be called a 'review' of the infamous Ovilus. That's actually a string of posts, but the latest has gotten more attention than the others overall. I'm not certain that the Ovilus post is popular because people agree with my point of view, but if the commentary following it is any indication, many do not. And that's ok, anyone and everyone is entitled to disagree with me. I'm generally quite willing to debate the issues and make my own case, and at the same time, I'll happily listen to your well-reasoned arguments (providing they are well-reasoned, on topic and not personal attacks).

Another post that was popular is my critique of EVP, or electronic voice phenomenon. I wasn't exactly charitable with that topic, though I didn't think I was being particularly harsh either. I simply said that the evidence gleaned through so-called EVP research isn't evidence of ghosts. Or rather that it can't be considered evidence of ghosts because there are too many other possible causes for the phenomenon. Though I acknowledge the skeptic explanations of pareidolia or even delusion – which, although it can seem like a dismissive and insulting suggestion, is a good possibility in some cases – I'm actually talking about other environmental causes, like unexpected electromagnetic fields and their effect on recording mediums, as only one example.

Today's post is connected to that very idea. Among the tool kit of the modern Ghost Hunter there are numerous pieces of equipment that are considered staples of the endeavour. Flashlights, cameras, micro-recorders and maybe even the Ovilus (much to my own chagrin). But the one device that nearly every Ghost Hunter relies on in their investigations is the EMF meter, and I question this practice in much the same way as I question the collection of EVP data.

The EMF meter is, for those not already familiar, a device for measuring electromagnetic fields in the general proximity of the device's sensors. Electromagnetic fields are everywhere, quite literally. Our environment, that is the entire planet, is inundated with waves of energy. The most powerful come from our sun, most of which is absorbed by our atmosphere, but everything that uses electricity in any way also generates an EM field. Electromagnetic energy is a form of radiation and it is measured along a spectrum – the electromagnetic spectrum – which includes at its extreme ends: Gamma radiation and ELF waves (extremely low frequency). Along the spectrum between Gamma rays and ELF are frequencies that correspond to radio waves (which includes cellular signals and other communications transmissions) and even the light spectrum that we're able to detect with our eyes (the visible light spectrum).

EMF meters are generally designed to measure AC currents that are generated by electric appliances with frequencies between 50 and 60 Hz, with obvious variation between different models, but they can all detect EM fields outside of those generated by AC current.

Now, to be perfectly clear, I'm not qualified to argue about the technical specifications of EMF meters, nor am I an expert on electromagnetic theory, but I do have a working understanding. I submit though, that such knowledge isn't necessary to see the flaw in this situation. I'll explain.

It's really as simple as this: there are many possible causes of EMF fluctuation in a given space, and as long as there is doubt about what caused the reading, it cannot be attributed to any one phenomenon.

The more experienced among you will already be gearing up to type out a comment pointing out that you, specifically aren't claiming that these readings are attributed to one phenomenon and/or that baseline readings are always taken (though I doubt the word "always" applies) to differentiate those readings from what are commonly called emergent EM fields (that would be an unexpected field of electromagnetic energy that doesn't appear to have been caused by sources that contributed to the baseline reading. Unexpected and unexplained.) If that's you, the one getting ready for a fight, relax, I'm not referring to you.

With some exception, the Ghost Hunter will claim that this emergent field is evidence of the presence of a ghost, or more reasonably evidence of a paranormal event. It is not.

Well, let me hedge my bets a bit here; it probably is not. Even with the utmost care taken to minimize the influence of electrical appliances in the area, there are errant EMF fields everywhere, fluctuating, cancelling each other out and permeating every structure and every living body. They're there, and they can be measured by an EMF meter. Cheaper meters are much more susceptible to interference from errant EMF than are the more expensive meters, which generally work in a different way than say, the K2 meter.

But all of this is beside the point. Unless the source of the field can be identified, it cannot be attributed to any one phenomenon. But I repeat myself.

Much of this boils down to a problem the paranormal community is dealing with in a most ineffectual way. The question is, why do Ghost Hunters use EMF meters in their investigations? I've asked around, and I got some weird answers. Some talked about the likelihood that an unexplained field was direct evidence of the paranormal; basically saying that if baseline sources are ruled out, it must be paranormal. Others said, indirectly, that they do it because someone else does (not in those words of course, but when one says that they don't understand it well enough to explain it, so go check so-and-so's blog for an answer, the logical inference is that they only do it because so-and-so made them think it was a good idea).

And this leads me to the real point: having not had the opportunity to survey the entire community of Ghost Hunters around the world, I'm stuck with generalizations, but from what I have seen and the conversations that I've had, a large portion of that community really doesn't understand what it is they are measuring with an EMF meter. Many use models that don't even given them an actual EMF reading, they just have lights that turn on when a field is detected. What, if anything, does this tell you about what is causing the field?

The presence of an unexpected or unexplained electromagnetic field is an anomaly. It may or may not be relevant to the investigation at hand, but in order to determine its relevance, one must identify its source. It answers no questions to simply squeal and declare that there be ghosts here.

From a scientific standpoint, there may be some merit to the idea that certain phenomena can be identified and/or tracked using a measurement of electromagnetic energy. However, for that endeavour to bear fruit, the researcher must have both an understanding of what the meter is measuring, how it measures it and what that measurement means in relation to every element of the environment in which it was observed. Pointing your meter into a dark room and declaring that any readings you get are the result of ghosts is quite silly.

So, I said it. I meant it. And I fully expect that many will disagree. I know that there are also quite a few who agree with me, and I cordially invite both groups to voice their opinion on the matter in the comment section below. If you don't wish to enter the argument, then I ask that you simply ask yourself, if you use an EMF meter to search for ghosts or for paranormal activity: why?

# The Doppelgänger and His Evil Twin

How would you react to seeing an exact duplicate of someone you love? Would you be able to tell which was which? Who is the real one and who is the...imposter?

What if you encountered that imposter independently? What if they not only looked like, but behaved and sounded like the original?

This is, apparently, exactly what happened to reddit user zeejoo12, who describes a bizarre encounter with what some have called a real life Doppelgänger. He posted his account of a strange mix up with his current girlfriend, wherein she appears at his apartment and confronts him for cheating on her, apparently. She proceeds to slap him and break several personal items and then storms away in a taxi cab. But to the man's surprise, his real girlfriend appears out of nowhere, claiming to know nothing of the blow-up.

Confused and frightened, the man turned to the reddit community for help in understanding exactly what had happened. His full story can be found on the sub-reddit: /r/Glitch_in_the_Matrix (18 June, 2013).

Despite the odd nature of that account, and the obvious opportunity for hoax, there are some elements of the story that intrigue. There may be more questions than answers and many of them are explored in the epic trail of comments and speculation that follow the original post, but much of this tale conforms to the traditional notion of a Doppelgänger.

The word Doppelgänger is German for double-goer and was coined by 18th century author Jean Paul (born Johann Paul Friedrich Richter) in his 1796 book Siebenkäs. But the concept of the supernatural double dates back as far as Ancient Egyptian mythology (ka) and Norse mythology (vardøger), and possibly earlier. Traditionally, a Doppelgänger is thought to be the paranormal double of a living person; an exact copy with memories and interpersonal connections intact.

The concept of the Doppelgänger is closely related to the metaphysical concept of bilocation, an idea with religious undertones and ties to many different philosophical systems; from shamanism and paganism, to Buddhism and Christian mysticism. Bilocation, however, is actually the appearance of one individual in two (or more) different places at the same time, rather than two examples of a single person existing at the same time.

The famous story of the English metaphysical poet John Donne recounts his experience with his wife's Doppelgänger on the night of his daughter's birth (stillbirth). *"I have seen a dreadful Vision since I saw you: I have seen my dear wife pass twice by me through this room, with her hair hanging about her shoulders, and a dead child in her arms: this, I have seen since I saw you."*[1]

Minus the dead child, Donne's account seems eerily similar to the story told by zeejoo12.

Other famous accounts of Doppelgänger encounters, like that of Johann Wolfgang von Goethe and even Abraham Lincoln – who saw himself in a mirror sporting two faces[2] – serve to solidify the paranormal nature of the phenomenon, and since then countless fictional works have been dedicated to the notion of a dark double.

Today the term doppelganger (as opposed to the traditional Doppelgänger) is commonly used to describe people who simply look alike, and with the prospect of human cloning looming in the not so distant future, it's likely that it will see greater use in contemporary culture.

Depending on which definition you choose, there are varying explanations for the Doppelgänger phenomenon. With our global population approaching more than 7 billion, the notion of more than one person looking eerily similar to another becomes a matter of statistics, rather than supernatural influence. But in those cases, like that above, where the experiencer(s) encountered the same person in two distinctly different locations or situations simultaneously, the prospect of explaining things gets a bit muddied.

Some have proffered ideas about mirror worlds or alternate dimensions briefly converging, creating unique circumstances that culminate in alternate realities clashing and interacting with each other. And without delving into yet another tangent explaining that dimensions are simply directions of travel and not places, suffice it to say, the semantics of the argument are less important than the foundation.

Interestingly, the many-worlds theory of quantum physics, or the many-worlds interpretation may actually support this idea. The many-worlds theory says that through universal wavefunction, all possible histories and futures exist simultaneously. This is different from the multiverse theory, as the latter is the assertion that outside of our physical universe exist many other universes, possibly with differing laws of physics.

Though the many-worlds theory doesn't say much about where those alternate histories and futures may be, physically. They could be overlaid on top of our reality or they could be separated by the infinite recesses of whatever exists beyond existence. In the case of the former, there do seem to be opportunities for such overlap to result in occasional temporary convergence, but any suggestion in that regard is pure speculation.

Other theories connect the Doppelgänger phenomenon with spirits and demons, claiming that the double is an evil incarnation of the original person, an evil twin if you will. The ancient mythological creatures incubus and succubus are sometimes connected to the Doppelgänger phenomenon, suggesting that they might take the form of their victim, or people close to them in their seductive efforts.

Somewhat lesser known are hypotheses involving time-slips, wherein the experiencer is thought to be viewing events that occurred in the past or will occur in the future. Which, surprisingly enough, is something Einstein showed is possible, if not technologically challenging.

You might think that the long and illustrious history of the Doppelgänger should be better understood than it actually is, but in truth, the phenomenon's transient nature coupled with its unpredictability have severely limited opportunities for investigation.  All that is available to researchers are after-the-fact accounts, often decades or even centuries old. Nonetheless, it can hardly be said that all the accounts are hoaxes or misidentifications, leaving the field wide open for further inquiry.

[1] Walton, Izaak. Life of Dr. John Donne. Fourth edition, 1675

[2] Sandburg, Carl. Abraham Lincoln: The Prairie Years. Harcourt, Brace and Co., New York, 1926. Volume 2, Chapter 165, pp.423-4

## The Many Faces of the Derenberger Incident

In the world of UFOlogy and Cryptozoology few characters are as spooky and memorable as those associated with the Derenberger incident.  And while many people may be unfamiliar with the case itself, the characters involved are infamous, both among paranormal circles and in popular culture.

The story of Woodrow Derenberger, a salesman for a small sewing machine company, has been told many times – most notably in the book he coauthored with Harold W. Hubbard called Visitors from Lanulos[1], which chronicled this strange tale.

It began on November 2, 1966, Derenberger was returning from a trip to Marietta, Ohio on his way home to Mineralwells, West Virginia (also Mineral Wells).  Derenberger was travelling alone, driving along interstate 77 near Parkersburg, West Virginia, when he was suddenly overtaken by another vehicle.  This vehicle, which Derenberger described as the strangest thing he'd ever seen – claiming it resembled a huge "kerosene lamp" – passed by Denereberger's truck and turned sideways, blocking both lanes of the highway, causing both vehicles to come to a complete stop.

Stunned, Derenberger watched as a man emerged from the strange vehicle.  The dark suited and oddly grinning man approached Derenberger's door as the strange vehicle lifted off the ground and floated to about 40 feet off the ground.

The man communicated with Derenberger through some form of telepathy, and identified himself as a seeker.  He asked Derenberger who he was and then told the frightened salesman that his name was Cold.  Cold asked several questions of Derenberger, some pertaining to the nearby town of Parkersburg, and after a short while, he ended the conversation by saying "It's been nice talking to you Mr. Derenberger.  We will be seeing you again."  And with that, Cold stepped away from the truck, at which point his strange vehicle returned to the roadway, and shooting off into the sky, he left the bewildered man sitting alone on the highway.

Just ten days later, the now famous events of the Mothman Incidents in Point Pleasant, West Virginia began, which apparently culminated in the collapse of the Silver Bridge in Point Pleasant, killing 46 people. As has been popularised by Mark Pellington's 2002 thriller The Mothman Prophecies staring Richard Gere and based on John Keel's book of the same name, Indrid Cold was a key figure in those events. Connecting that Cold with the one in Derenberger's version, some suggest that Cold was (or is) a member of the elusive and mysterious Men In Black, while others believe he is an alien or inter-dimensional being.

But his story doesn't end there, and neither does Woodrow Derenberger's. Following the events of Point Pleasant, Derenberger made claims that Cold had admitted, through continued telepathic contact, that he was an alien from a planet called Lanulos within the galaxy known as Genemedes (both of which seem to be fictional). More than that, however, Derenberger claimed that Cold had actually taken him to Lanulos in a spaceship, where Derenberger claimed to have seen many other Lanulosians and relayed some commentary on their culture. Over the years, according to Derenberger, Cold was joined on earth by two other Lanulosians, named Demo Hassan and Karl Ardo, both of whom were apparently more discreet than Cold ever was.

Wild as this may seem, some believe that these events are corroborated by their connection to another terrifying paranormal phenomenon: The Grinning Man. First mentioned by paranormal researcher and author John Keel (yes the same man who wrote Mothman Prophecies), in his seminal work The Complete Guide to Mysterious Beings [2], the Grinning Man, or men as the case may be, is a strange and rather large man who, like the name suggests, grins, and scares the bejesus out of those who see him.

Keel believed that the first ever encounter with The Grinning Man occurred in Elizabeth, New Jersey on October 11, 1966, less than a month before Derenberger's encounter. On that night two young boys, Martin Munov and James Yanchitis were walking home along a road that ran adjacent to the elevated New Jersey Turnpike. As they walked along the dark street, Yanchitis noticed a strange looking man standing in the darkness at the top of the treacherous incline, trapped by a chain link fence. He then called out to Munov, and the boys watched as the figure slowly turned to face them with an unnerving ear to ear grin.

The boys described the being as taller than six foot two inches (referencing one of the investigators of the case, the famous actor Chuck McCann, who was a rather large individual) and very broad. They said that they saw no facial features other than the creature's beady eyes and its wide toothy grin. They also noticed that the being was wearing a kind of shimmery green jumpsuit or overalls.

Munov and Yanchitis' story was documented by the police, the press and by Keele and his team of investigators, and other accounts of a strange tall, grinning man began to come in from all around the area. Keele connected this Grinning Man encounter with an alleged UFO sighting that occurred at the same time some 40 miles north of Elizabeth, near the DuPont explosives factory outside Pompton Lakes, New Jersey. Those familiar with the Mothman incident will note that there were several Mothman sightings in an area of Point Pleasant known as the "TNT area", which had been a munitions manufacturing and storage area previously.

As with many stories of this nature, the connections between incidents are tenuous at times and some point out that since Keel was almost exclusively the only one to document these cases, he may have used poetic license in his tellings of the stories, drawing connections where they might not have existed in reality. Because of the relatively close proximity of the encounters and the timeframe in question, many believe that Keele was justified in his speculation that the Grinning Man and Indrid Cold were one and the same. Cold's apparently prophetic involvement in the Point Pleasant disaster paints a picture of the creature, or whatever you might call him, as something to be feared, and if the Grinning Man encounters are indeed connected to Indrid Cold, perhaps there's something to be said for that hypothesis. But Woodrow Derenberger's accounting suggests that Cold and his associates are/were simply curious travellers.

Unfortunately, there have been no modern sightings, of Cold or the Grinning Man, so if there is any thread of truth weaved through this story it is likely to remain hidden into the foreseeable future. The stories of Indrid Cold and the Derenberger Incident are favourites of many in the paranormal community, and much speculation has been focused on these reports. Really the only source materials available are Keele and Derenberger's books, but that doesn't mean that there aren't many tellings of the stories online. Anyone looking for more information can easily find it with a simple search.

[1] Derenberger, Woodrow. Visitors from Lanulos. Vantage Press (1971)

[2] Keel, John. The Complete Guide to Mysterious Beings. Tor Books (2002). ISBN-10: 0765345862 Chapter 14. http://galaksija.com/literatura/guide.pdf

# The Cottingley Fairies

Of the silly Fortean beliefs held by people of the late 19th and early 20th century, some are more understandable than others. When an idea infects the populous it can spread as quickly as a scientifically engineered super-virus, and when those ideas are backed by the likes of intellectual giants such as Sir Arthur Conan Doyle, it's easy to forgive the people for buying into it.

I characterize Doyle as an intellectual giant, and he rightly was, but to be honest, he got behind some fairly weird ideas during his heyday. Doyle was a huge proponent of the séance, he held an immense fascination with tipping tables and the medium's trumpet and ectoplasm, but his attentions weren't only focused on the ghostly and the spiritual. In fact, his interest and passion for cryptozoological topics resulted in his being taken in by a few well executed hoaxes over the years.

One such example was the Cottingley Fairies. As the story goes, in mid-1917, two young girls, cousins Elise Wright and Frances Griffiths of West Yorkshire, England, visited Cottingley beck, a small stream running near their small village. There the girls apparently witnessed groups of fairies frolicking about near the river's edge. Elise being 16 at the time and Frances being only 10 years old, claimed to not only interact with the fairies but, incredibly, they captured five photographs of the creatures over a period of time.

    Showing several small (approximately six inch tall), apparently female fairy like creatures, the photos became public in mid-1918 through the Theosophical Society of Bradford England, via one of the Society's leading members, Mr. Edward Gardner.  Gardner subjected the photos to analysis, such as it was at the time, and with the help of photography expert Harold Snelling, determined that the photos were not faked.  At least as far as saying that the photos showed what had been presented to the camera at the time of exposure and were not manipulated photographically.

    Elise's Father, a professional photographer himself, with a darkroom set up in his home and whose camera the girls had borrowed to take the photos, had dismissed the photos as a hoax, believing the fairies to be cardboard cut-outs, but Frances' Mother was taken by the photos and was the one to bring them forward to Gardner.

    Public opinion was split over the authenticity of the photos and the girls enjoyed some short-lived celebrity over the incident.  They eventually became disenchanted with the idea of fairies and the attention of investigators and psychics interested in exploiting the situation soon became a burden to the girls and their families.

    Conan Doyle became aware of the photos through the editor of the Spiritualist publication Light.  And having been commissioned to write an article on fairies for the 1920

Christmas issue of The Strand Magazine, Doyle used the photos as the basis for his article, interpreting them as clear and visible evidence of psychic phenomenon.

Following Conan Doyle's involvement, further analysis took place via the photography companies Kodak and Ilford, and while the Ilford technicians found evidence that the photos had been faked, the Kodak analysis agreed with Snelling's initial assessment.

Like most incidents of paranormal phenomenon, the fervour and public interest in the photos eventually died down and Elise and Frances went on with their lives. Until, that is, the BBC covered the story in 1971 in their Nationwide programme in which Elise maintained her original story, that she believed the fairies were figments of her imagination and that she had somehow managed to photograph her own thoughts.

Over the years Elise and Frances had been interviewed for other programmes and news stories and they always maintained their story. James Randi, in cooperation with a team from the Committee for the Scientific Investigation of Claims of the Paranormal, concluded through computer analysis that strings could be seen suspending the fairies in mid-air. And in 1983 the cousins admitted in an article published in the magazine The Unexplained that they had in fact hoaxed the photos.

Elise admitted that the fairies were cardboard cut outs from a children's book (Princess Mary's Gift Book – 1914), and that they had been suspended in front of the camera using hat pins. While Elise claimed that all five photos were faked, Frances maintained, quite adamantly, that the fifth photo was genuine and that it did in fact depict real fairies that they saw at Cottingley beck.

The women, now deceased, were responsible for pulling the wool over the eyes of many a learned and experienced investigator, not the least of which was Conan Doyle, and their accounts of the events surrounding the photos never waivered, up until their admission of the hoax. It's a near certainty that some people still believe the fairies are real, in spite of the confession, and bearing that in mind, can one forgive the great genius of Conan Doyle for being duped by a couple of adolescent girls with vivid imaginations and the creative expertise to affect some of the most convincing fairy photos in history?

This wasn't the first time Conan Doyle was taken in by a story of incredible proportions and it certainly wasn't the last, but in my mind, his passion for the subject and the reach of his own imagination outweigh his ultimate gullibility.

# The Nazi Bell, Wunderwaffe or Time Portal?

Secret facilities, hidden deep in northern Germany, leaked and de-classified documents, anecdotal evidence and urban legend, all of it pointing in one direction. And what direction is that?

Die Glocke or in English, "The Bell" is believed to have been Nazi Germany's famed Wunderwaffe or Wonder Weapon. It was the culmination of Nazi Germany's brightest scientific minds. The same people who brought us the V1 and V2 rockets (the V2 being the first manmade object to leave our atmosphere and plunge into the cold depths of space) are also thought to have been involved in the development of a weapon so terrible that some accounts describe factions of these scientists refusing to release technical plans to Nazi leaders for fear of what might be done with the technology. Incidentally, these were also the men who eventually helped to design and build the first atomic bomb.

The Bell, however, is a complete mystery, though there is no shortage of conspiracy surrounding it. First word of Die Glocke was presented to the free world in 2000 by one Igor Witkowski, a polish author who wrote a book titled (in Polish) Prawda O Wunderwaffe or The Truth About The Wonder Weapon. Witkowski claimed that he met with an unnamed Polish Intelligence contact in 1997, from which he reports to have been shown classified Polish Government documents detailing Nazi weapons research projects. Die Glocke was one of those projects.

The story of Die Glocke was later picked up by British author Nick Cook who added his own flavour to the tale in his own book The Hunt for Zero Point. Whatever the truth about "The Bell", conspiracy theorists and even some in the general scientific community are convinced that the Bell did in fact exist and that it was a machine of incredible power.

For the record, no one really knows what Wunderwaffe actually was, or even if Nazi Germany was really developing anything other than conventional weaponry. Most mainstream science and historical experts are adamant that the V series rockets were the pinnacle of German technology at the time, and that the Bell is a simple urban legend. But I'm not about to leave it there…you knew I wouldn't.

Most who believe that the Bell exists, or existed at some point, are convinced that it was a machine built for a fantastic and sinister purpose. Many believe that it was either a working time machine or an antigravity machine. If you believe what you see on TV and read on the internet, the Third Reich was indeed undertaking some rather nefarious research and development just prior to the end of the war in 1945. That is, aside from their ballistic and chemical weapons programs.

Allegedly an experiment was carried out by Third Reich scientists working for the SS in a German facility known as Der Riese ("The Giant") near the Wenceslaus mine and close to the Czech border. Die Glocke is described as being a device "made out of a hard, heavy metal", approximately 9 feet wide and 12 to 15 feet high having a shape similar to that of a large bell. According to Cook, this device ostensibly contained two counter-rotating cylinders which would be "filled with a mercury-like substance, violet in color. This metallic liquid was code-named "Xerum 525" and was otherwise cautiously "stored in a tall thin thermos flask a meter high encased in lead". Additional substances said to be employed in the experiments, referred to as Leichtmetall (light metal), "included thorium and beryllium peroxides". Cook describes Die Glocke as emitting strong radiation when activated, an effect that supposedly led to the death of several unnamed scientists and various plant and animal test subjects. Based upon certain external indications, Witkowski speculates that the ruins of a metal framework in the vicinity of the Wenceslas mine (aesthetically dubbed "The Henge") may have once served as test rig for an experiment in "anti-gravity propulsion" generated with Die Glocke; others, however, dismiss the derelict structure as simply being a conventional industrial cooling tower.

If we allow our speculations to run wild, we can easily come up with any number of scenarios, from time travel, to manipulation of space-time, to antigravity and even to inter-dimensional travel. Some German scientists have gone on record stating that the machine was designed to warp space-time and to allow the SS to travel backward through time, though since the majority of the world still speaks English, we can safely say that their plan didn't work...or did it?

To anyone familiar with my work, you already know that I have a penchant for physics, cosmology and time travel, and I would like to put forth a theory of my own.

Dr. Brian Greene, author of The Fabric of the Cosmos, among others, has outlined a thought experiment, wherein one thinks of moments in time as slices of bread in a very long loaf. Each slice corresponds to the slice behind it and before it, but none other. If you walk along the loaf you can pluck out a slice from anywhere and that slice will be the present. All of the slices before it represent the future and all of the slices behind it are the past. This analogy specifies the direction of time's arrow, and it is only the beginning of my theory.

Travelling forward along the loaf is easy enough, each day, each minute each second of our lives we are doing just that, moving forward through time. Moving backward however, proves to be difficult. The scientific barriers to backward time travel or retro-time travel notwithstanding there is a paradox which must first be overcome before anyone can change time's arrow; it is the Grandfather Paradox.

The Grandfather Paradox says, quite simply, that one cannot travel into the past and kill one's Grandfather. The reason seems obvious enough, if the traveller kills his Grandfather, his Father will never be born. Hence the traveller will never be born in order to grow up, discover time travel and go back to kill his Grandfather.

The same paradox exists for nearly anything you could imagine doing in the past, any changes that might be made, which could have an impact on the life of the traveller, would serve to make retro-time travel an impossibility.

There are though, a few ways around this paradox, and they might just relate to the Nazi Bell after all.  First off, there is the idea that the past is unchangeable, so going back in time would have you living your life as you did originally during the time period in question, and were you to travel back beyond your own birth, you would cease to exist.  Second and more important if not poignant, there is the idea that travelling backward through time has you travelling to a parallel time, or in line with the loaf of bread analogy above, an alternate loaf.  You wouldn't actually be travelling back to a previous slice of bread from the past; you would be travelling to a corresponding slice of bread in a different loaf.

If this were the case, any changes you made would then play out in an all new and distinct future, thus, if you murdered your grandfather in that time line, only the "you" who was indigenous to that timeline would suffer for it.

And this finally brings me back to the Nazi's and their Wunderwaffe, if one were to subscribe to the idea that the Nazi's had succeeded in building a time machine, with the intent to travel backward in time, for the obvious peril of us all, one has first to consider the Grandfather Paradox.  Might it be possible that SS scientists succeeded, and were transported via Die Glocke to an alternate past, and in that parallel timeline are ruling over all mankind?

The above is really a discussion of the quantum physics theory called Many-Worlds, wherein every conceivable timeline exists parallel to our own.  It purports that every possible configuration of the universe exists and is equally as real as the one we inhabit.

So, did Nazi scientists develop and build a time machine?  Your guess is as good as mine, though, I have heard reports that several top SS Officers and scientists disappeared without a trace just prior to the end of the war...

www.ingramcontent.com/pod-product-compliance
Lightning Source LLC
Chambersburg PA
CBHW040904180526
45159CB00010BA/2916